ENVIRONMENTAL RESEARCH ADVANCES

KITCHEN EXHAUST VENTILATION

ADOPTION AND CAPTURE EFFICIENCY ISSUES

Environmental Research Advances

Additional books in this series can be found on Nova's website under the Series tab.

Additional e-books in this series can be found on Nova's website under the e-book tab.

ENVIRONMENTAL RESEARCH ADVANCES

KITCHEN EXHAUST VENTILATION

ADOPTION AND CAPTURE EFFICIENCY ISSUES

MALLORY P. MICHAELS
EDITOR

NOVINKA

New York

For permission to use material from this book please contact us:
Telephone 631-231-7269; Fax 631-231-8175
Web Site: http://www.novapublishers.com

Library of Congress Cataloging-in-Publication Data

ISBN: 978-1-63463-642-1

Published by Nova Science Publishers, Inc. † New York

CONTENTS

Preface		**vii**
Chapter 1	Addressing Kitchen Contaminants for Healthy, Low-Energy Homes *J. Chris Stratton and Brett C. Singer*	**1**
Chapter 2	Performance of Installed Cooking Exhaust Devices *Brett C. Singer, William W. Delp, Michael G. Apte and Philip N. Price*	**57**
Chapter 3	Pollutant Removal Efficiency of Residential Cooking Exhaust Hoods *Brett C. Singer, Alexander D. Sherman, Toshifumi Hotchi and Douglas P. Sullivan*	**79**
Index		**89**

PREFACE

Chapter 1 - Cooking and cooking burners emit pollutants that can adversely affect indoor air quality in residences and significantly impact occupant health. Effective kitchen exhaust ventilation can reduce exposure to cooking-related air pollutants as an enabling step to healthier, low-energy homes. This report identifies barriers to the widespread adoption of kitchen exhaust ventilation technologies and practice and proposes a suite of strategies to overcome these barriers. The recommendations have been vetted by a group of industry, regulatory, health, and research experts and stakeholders who convened for two web-based meetings and provided input and feedback to early drafts of this document. The most fundamental barriers are (1) the common misconception, based on a sensory perception of risk, that kitchen exhaust when cooking is unnecessary and (2) the lack of a code requirement for kitchen ventilation in most US locations. Highest priority objectives include the following: (1) Raise awareness among the public and the building industry of the need to install and routinely use kitchen ventilation; (2) Incorporate kitchen exhaust ventilation as a requirement of building codes and improve the mechanisms for code enforcement; (3) Provide best practice product and use-behavior guidance to ventilation equipment purchasers and installers, and; (4) Develop test methods and performance targets to advance development of high performance products. A specific, urgent need is the development of an over-the-range microwave that meets the airflow and sound requirements of ASHRAE Standard 62.2.

Chapter 2 - The performance metrics of airflow, sound, and combustion product capture efficiency (CE) were measured for a convenience sample of fifteen cooking exhaust devices, as installed in residences. Results were analyzed to quantify the impact of various device- and installation-dependent parameters on CE. Measured maximum airflows were 70% or lower than values noted on product literature for 10 of the devices. Above-the-cooktop devices with flat bottom surfaces (no capture hood) – including exhaust fan/microwave combination appliances – were found to have much lower CE at similar flow rates, compared to devices with capture hoods. For almost all exhaust devices and especially for rear-mounted downdraft exhaust and microwaves, CE was substantially higher for back compared with front burner use. Flow rate, and the extent to which the exhaust device extends over the burners that are in use, also had a large effect on CE. A flow rate of 95 liters per second (200 cubic feet per minute) was necessary, but not sufficient, to attain capture efficiency in excess of 75% for the front burners. A-weighted sound levels in kitchens exceeded 56 dB when operating at the highest fan setting for all 14 devices evaluated for sound performance.

Chapter 3 - Capture efficiency (CE) of exhaust from a natural gas cooking range was quantified for three common designs of residential range hoods in laboratory experiments: (A) microwave exhaust combination; (B) short hood with grease-screen-covered air inlet at bottom; and (C) deep, open hood exhausting at top. Devices were evaluated at varying installation heights, at highest and lowest fan settings, and with the hood installed 15 cm away from back wall with intent to improve CE for front burners. Each configuration was evaluated for the oven and for three cooktop burner combinations (two back, two front, one front and one back). At highest fan settings and standard installation against the wall, Hoods A and C captured back cooktop burner exhaust at >90% and Hood B at >80%. In this configuration, CE for front burner exhaust was 73-78% for Hoods A and C but only 46-63% for Hood B. CEs followed similar patterns but were substantially lower on the lowest fan speed. Installing the hood away from the wall improved CE for oven and front burners on Hood A at low speed, but substantially reduced CE for back burners for all hoods at low and high speed.

In: Kitchen Exhaust Ventilation
Editor: Mallory P. Michaels

ISBN: 978-1-63463-642-1

Chapter 1

ADDRESSING KITCHEN CONTAMINANTS FOR HEALTHY, LOW-ENERGY HOMES*

J. Chris Stratton and Brett C. Singer

ABSTRACT

Cooking and cooking burners emit pollutants that can adversely affect indoor air quality in residences and significantly impact occupant health. Effective kitchen exhaust ventilation can reduce exposure to cooking-related air pollutants as an enabling step to healthier, low-energy homes. This report identifies barriers to the widespread adoption of kitchen exhaust ventilation technologies and practice and proposes a suite of strategies to overcome these barriers. The recommendations have been vetted by a group of industry, regulatory, health, and research experts and stakeholders who convened for two web-based meetings and provided input and feedback to early drafts of this document. The most fundamental barriers are (1) the common misconception, based on a sensory perception of risk, that kitchen exhaust when cooking is unnecessary and (2) the lack of a code requirement for kitchen ventilation in most US locations. Highest priority objectives include the following: (1) Raise awareness among the public and the building industry of the need to install and routinely use kitchen ventilation; (2) Incorporate kitchen exhaust ventilation as a requirement of building codes and improve the mechanisms for code enforcement; (3) Provide best practice

* This is an edited, reformatted and augmented version of a report, LBNL-6547E, issued by the Lawrence Berkeley National Laboratory, January 2014.

product and use-behavior guidance to ventilation equipment purchasers and installers, and; (4) Develop test methods and performance targets to advance development of high performance products. A specific, urgent need is the development of an over-the-range microwave that meets the airflow and sound requirements of ASHRAE Standard 62.2.

EXECUTIVE SUMMARY

Cooking and cooking burners can emit combustion products, pollutants and excess moisture that may adversely impact indoor air quality in residences. Cooking-related contaminants include but are not limited to carbon monoxide, nitrogen dioxide, formaldehyde, volatile organic compounds, polycyclic aromatic hydrocarbons, fine particulate matter, and ultrafine particles. People present in a home during cooking can perceive odors, smoke, and dampness, but the level of concern about these cooking-related contaminants varies. People cannot directly detect or assess the quantity of pollutants in the air and many are unaware that cooking and cooking burners produce health-relevant pollutants. As a result, use of kitchen ventilation is inconsistent even when it is available, and efficacy is not a strong market driver for kitchen designs or ventilation products.

Recent research indicates that cooking-related pollutants have a greater impact on health, safety and indoor air quality than had previously been recognized, and for this reason should now be given more serious attention.

Few US states have mandatory codes or standards relating to kitchen ventilation. As a result, many homes have been built and are being built today without any kitchen ventilation. Even in homes that have kitchen ventilation, the systems may not be very effective owing to a variety of factors. Capture efficiencies of 100% are possible with a well-designed exhaust hood that has adequate airflow and is positioned above a cooktop burner or oven exhaust. Efficiency is lower for systems with inadequate airflow or exhaust hoods that are not directly above the burners. An exhaust fan at another location in the kitchen is inherently not as efficient at capturing pollutants before they mix into the breathing space. High resistance ductwork works to reduce airflow for many exhaust fans. Exhaust hood design can also play a role. Current industry standards (e.g., ASHRAE 62.2) focus on airflow, not efficacy at capturing and removing pollutants. The application of best practice to kitchen ventilation is extremely limited.

We developed the assessment and recommendations outlined in this document in consideration of available information, input from stakeholders and experts that participated in webinars convened in May and June of 2013, and comments by reviewers of early drafts of the document. Recommendations are presented in two sections, broadly organized around Market Transformation and Technology and Product Development. Recommendations are organized into Objectives deemed needed to bring about a transition to reduced kitchen pollutant exposure in homes and Objectives are linked to specific Actions; Research , Development and Demonstration needs; and a Technology Development roadmap.

We highlight the following four activities as cornerstones to reducing the impact of cooking related pollutants in homes:

1) Raise awareness among the public and the building industry of the need to install and routinely use kitchen ventilation;
2) Incorporate kitchen exhaust ventilation as a requirement of building codes and improve the mechanisms for code enforcement;
3) Provide best practice product and use-behavior guidance to ventilation equipment purchasers and installers, and;
4) Develop test methods and performance targets to advance development of high performance products.

Developing a test method for pollutant removal effectiveness is a specific high-priority, near-term objective. Over the mid- to longer-term, the automation of kitchen exhaust or alternative pollutant removal mechanism(s) is a key technology target to ensure robust control. Incentivizing or requiring automated range hoods in codes in standards would serve to propagate this functionality into "base" range hood models and into operation in homes. There is also an urgent need for one or more manufacturers to develop over-the-range (OTR) microwaves that meet the airflow and sound requirements of ASHRAE 62.2. The lack of any ASHRAE 62.2-compliant OTR microwaves creates a tension for homebuilders and contractors that could easily be resolved by industry product development.

I. Market Transformation Objectives

O1. Expand awareness of the importance of kitchen ventilation.
O2. Expand the use of kitchen ventilation.

O3. Publicize existing guidance for building professionals on the basics of kitchen ventilation and develop guidance on best practice designs and equipment specifications.

O4. Provide better performance information and guidance to purchasers and users of kitchen ventilation products.

O5. Improve and expand accessibility of guidance for installers and inspectors.

O6. Raise awareness of kitchen ventilation requirements of ASHRAE Standard 62.2 (Ventilation and Acceptable Indoor Air Quality in Residential Buildings).

O7. Establish capture efficiency as a key performance measure for range hoods.

O8. Promote and support development of automatic operation technologies for kitchen ventilation.

O9. Develop tools and guidance to expand commissioning of kitchen ventilation systems.

O10. Reduce the number of new homes built without vented kitchen exhaust.

O11. Incorporate effective kitchen ventilation as a standard element of energy retrofits.

O12. Develop and demonstrate kitchen ventilation systems for low- and zero-net-energy, high performance homes.

Specific Actions

A1. Make kitchen ventilation required in the International Residential Code (IRC), International Mechanical Code (IMC), International Energy Conservation Code (IECC), International Green Construction Code (IgCC), Uniform Mechanical Code (UMC), and mechanical codes at state and local levels.

A2. Increase awareness of airflow rating certification; ensure that best practice guides and home performance standards specify airflows certified at an appropriate static pressure.

A3. Facilitate development and adoption of standard method of test for capture efficiency (pollutant removal effectiveness).

A4. Incorporate capture efficiency into standards including ASHRAE 62.2, ENERGY STAR label for range hood products, and high performance home standards.

A5. Align Home Ventilating Institute (HVI) airflow test and sound procedures and ASHRAE standard 62.2 prescriptive compliance pathway to achieve effective performance of installed range hoods.
A6. Incorporate automatic operation into best practice guides and high performance home standards as products are determined to be reliable based on standard test method.
A7. Incorporate into combustion safety diagnostics a consideration of whether the range hood has an automatic timer shut-off feature.

Research, Development, and Demonstration

R1. Develop and demonstrate effective methods for commissioning range hoods.
R2. Characterize in-field use and performance of range hoods.
R3. Develop a comprehensive contaminant removal effectiveness metric.
R4. Improve or validate ASHRAE 62.2 "deficit make-up" approach for existing buildings.
R5. Evaluate reasonable combustion safety allowance for timed auto-off kitchen ventilation.
R6. Develop standard test method and performance metrics for automatic operation.
R7. Re-evaluate kitchen ventilation and moisture removal.
R8. Evaluate potential for effective unvented range hood.

II. Technology and Product Development Roadmap

T1. Develop ASHRAE 62.2-compliant microwave range hoods.
T2. Expand the selection of quiet, effective, affordable and energy-efficient range hoods.
T3. Establish automatic range hood shut-off to reduce hazard associated with natural draft combustion appliance backdrafting.
T4. Integrate kitchen exhaust into smart ventilation systems for residences.
T5. Ensure performance through on-board diagnostics.
T6. Develop effective range hoods with automatic operation.
T7. Establish effective low-energy kitchen ventilation options.

ACKNOWLEDGMENTS

Support for this work was provided by the U.S. Dept. of Energy Building America Program, Office of Energy Efficiency and Renewable Energy under DOE Contract DE-AC02-05CH11231; by the U.S. Dept. of Housing and Urban Development, Office of Healthy Homes and Lead Hazard Control through Interagency Agreement I-PHI-01070; by the U.S. Environmental Protection Agency Indoor Environments Division through Interagency Agreement DW-89- 92322201-0; and by the California Energy Commission through Contract 500-09-042. We thank all those who contributed via participation in the webinars and/or reviewing early drafts of this document. We are particularly grateful to the thoughtful and detailed comments and suggestions provided by Paul Francisco, Dean Gamble, Peggy Jenkins, Mike Moore, Max Sherman, Peter Strait, Iain Walker, and Zoe Zhang.

PURPOSE AND SCOPE

The purpose of this report is to develop an agenda to achieve high performance kitchen ventilation or alternative solutions that will significantly reduce the potential negative indoor air quality impacts of cooking-generated contaminants while simultaneously enabling reductions in residential sector energy use. This agenda identifies needs and opportunities for advances through the mechanisms of policy, technology development, public awareness, and building industry tools and training. This report combines a review of the current state of the art of kitchen ventilation with input from industry stakeholders. The focus is on common residential cooking equipment and mitigation technologies suitable for use in new construction, retrofit, and product replacement in existing homes.

BACKGROUND

The Purpose of Kitchen Ventilation

The purpose of kitchen exhaust ventilation is to remove pollutants, moisture, smoke and odors generated during cooking. A good overview of the need for kitchen ventilation is provided by Parrott et al. (2003).

Pollutant emissions from cooking burners and the cooking of food can substantially and adversely impact air quality in homes. Natural gas burners commonly emit nitrogen dioxide (NO_2) and under some conditions emit substantial quantities of carbon monoxide (CO), formaldehyde (HCHO) and ultrafine particles (UFP) (Wallace et al. 2004; Singer et al. 2010; Dennekamp et al. 2001; Moschandreas et al. 1986; Moschandreas and Relwani 1989). Studies have shown associations between increased exposures to NO_2 in homes and increased respiratory symptoms in children including chest tightness, shortness of breath, wheeze, and increased number of asthma attacks (Belanger et al. 2006; Hansel et al. 2008; Garrett et al. 1998). Electric coil resistance burners produce UFP (Dennekamp et al. 2001). Cooking activities produce fine and ultrafine particles and a wide range of irritant and other potentially harmful gases including acrolein and polycyclic aromatic hydrocarbons (Abdullahi et al. 2013; Fortmann et al. 2001; Buonanno et al. 2009; Fullana et al. 2004; Seaman et al. 2009; Zhang et al. 2010). Gas burners and cooking can also release substantial quantities of water vapor that could contribute to moisture-related indoor air quality problems (Parrott et al. 2003). Individual cooking events can produce short-term PM2.5 concentrations exceeding 300 $\mu g/m^3$ and UFP concentrations exceeding $10^5/cm^3$ (Wallace et al. 2004; Abdullahi et al. 2013; Buonanno et al. 2009; Zhang et al. 2010; He et al. 2004; Afshari et al. 2005; Booth and Betts 2004; See and Balasubramanian 2008).

Kitchen exhaust ventilation can be provided via any of the following designs: a range hood or other exhaust device – including a combination microwave range hood – mounted above the cooktop and oven; a downdraft exhaust system mounted alongside the cooktop burners; an exhaust fan in the room containing the kitchen; an exhaust fan elsewhere in the home. Since the goal of kitchen exhaust ventilation is to remove moisture and pollutants from the location within the house where they are being released, kitchen ventilation is most effective when it is closest to moisture and pollutant sources. A range hood, downdraft exhaust vent and potentially even a well-placed wall or ceiling exhaust fan can be much more effective than an exhaust fan placed elsewhere in the kitchen or home.

The functions of pollutant, smoke, and odor removal theoretically can be provided by filtration and air cleaning equipment included in a recirculating range hood, provided through a local kitchen air cleaning device, or incorporated into a household central system. While there are many recirculating range hoods on the market that claim to be effective at removing smoke and odors, and some that claim to remove some pollutants, we know of no residential product that claims to remove CO or water vapor. There is significant doubt concerning recirculating range hoods' efficacy of removal for all these components and pollutants.

Factors Determining Effectiveness

The following elements impact the effectiveness of a kitchen ventilation system:

- ***System geometry and equipment type.*** The location of the exhaust device – above the cooktop, a downdraft device next to the cooktop, an exhaust fan above the range or wall-mounted oven, or an exhaust fan elsewhere in the kitchen or home – is a key determinant of potential effectiveness. Equipment can operate continuously or intermittently and can be automatic or require manual initiation.
- ***Equipment design and performance characteristics.*** Exhaust ventilation equipment varies widely in design and performance, with factors such as exhaust flow rate, hood volume, and burner coverage all impacting plume capture. Range hoods come in widely varying sizes and shapes: some include horizontally and vertically deep bottom areas that create large volumes where a plume may collect (the sump), whereas others have flat or curved bottoms – often with grease screens arrayed across the bottom inlet – and very limited or no sump volume. Exhaust fans vary in their maximum airflow capacity and their effectiveness in moving sufficient air, particularly with increasing airflow resistance of system ducting. The different morphologies and mechanical performance characteristics result in varying potential effectiveness at capturing pollutants, moisture and grease generated at the cooktop or in an oven beneath the cooktop.
- ***Exhaust ducting.*** The ducting beyond the exhaust fan sets the system airflow resistance characteristics that establish the pressure at the exhaust fan at a given airflow. Duct performance is impacted by the material (e.g., smooth or accordion flex), length, duct diameter, number of turns and transition pieces, and the exterior termination fitting. In general, higher pressure means higher noise levels and increased fan power required to produce a given airflow.
- ***Installation details.*** Installation details including fan energy consumption, pollutant removal effectiveness, and noise level can

substantially impact performance. For range hoods, the height above the cooktop and whether the cooktop and hood are in an area with wall cabinets, on an open wall, or on an island are important to the flow patterns that develop between the cooktop and hood. These characteristics should impact performance of wall and ceiling exhaust fans above a cooktop as well.

- ***Use patterns.*** It should be obvious that any system that is not used is ineffective. Conversely, even a partially effective system that is used all the time will substantially decrease pollution exposures in the home. Use behavior may be significantly impacted by noise.

The elements noted above have important interactions. For example, if the exhaust duct system is designed or installed in a manner that produces high air flow resistance, it is more important to select an exhaust device that operates effectively while connected to ducts with high air flow resistance. High airflow resistance will increase noise, which tends to decrease use and therefore efficacy.

Kitchen Exhaust Ventilation Currently in U.S. Homes

We are unaware of any published data that provide statistics on the current presence of kitchen ventilation systems in U.S. homes. Like many building features, the inclusion of kitchen ventilation into homes is dictated by code requirements and market-driven building practices. The major international building codes, on which many U.S. state codes are based, do not require kitchen exhaust ventilation. Since 2006, 2008, and 2010[1] respectively, Oregon, Washington, and California have required kitchen ventilation in their state building codes. Other states may have requirements and, even within states that do not require it, in some areas the inclusion of kitchen ventilation may be an established building practice. Webinar participants confirmed our understanding that there are many states and areas of the U.S. in which kitchen exhaust ventilation is rare.

A 2011 study (Klug et al. 2011a) documented range hood characteristics in California homes using data collected from a real estate web site. This study examined 1002 homes and found for example that 7.4% of the homes had no range hood installed and 47% of the homes had a microwave range hood installed[2]. This method did not include some important performance characteristics, such as airflow rate or whether the range hood is vented to the outside. Nevertheless, this low-cost data collection approach provides limited but valuable information about the presence and types of kitchen ventilation in homes, and could be expanded to examine range hood characteristics in homes across the US.

Relevant Codes, Standards, and Guidelines

ASHRAE Standard 62.2: Ventilation and Acceptable Indoor Air Quality in Low-Rise Residential Buildings

In the US, ASHRAE Standard 62.2[3] is the most important standard with kitchen ventilation requirements. The 2010 and 2013 versions of this standard require the following for kitchen ventilation:

For intermittent fans:

- Sound must be HVI rated for 3 sones or less at 100 cubic feet per minute (cfm);
- To use the prescriptive duct sizing compliance pathway[4], duct system must meet dimension requirements and fan must be HVI rated for 100 cfm at 62.5 Pa (0.25 IWC) of static pressure; otherwise
- Fan airflow must be measured as installed to move at least 100 cfm;
- If the fan's airflow is equal to or greater than 5 kitchen air changes per hour (ACH), then the fan may be a local kitchen exhaust fan or a range hood;
- If the fan's airflow is less than 5 kitchen ACH, the fan is required to be a range hood

For continuous fans:

- Must be HVI rated or measured as installed to move an airflow that corresponds to 5 kitchen ACH with a rated sound level of 1 sone or less

The procedure is straightforward for determining the airflow rate to provide 5 ACH for a kitchen that is a distinct room[5]. For kitchens in open floor plan homes without walls separating the cooking space from adjacent living space, the procedure is more ambiguous; required airflow rate depends on how the kitchen volume is calculated, and ASHRAE 62.2 does not include any clarifying language stipulating how kitchen volume shall be calculated in open floor plan homes. If the kitchen is interpreted to end where the cabinets end[6], the airflow to provide 5 kitchen ACH is significantly lower than if the kitchen volume is equal to that of the large open area of which it is a part.

There is currently a change proposal for ASHRAE 62.2 that would require 300 cfm of kitchen exhaust if the fan is not a range hood, 100 cfm of kitchen exhaust if it is a range hood, or 5 ACH of kitchen exhaust based on kitchen volume for an enclosed kitchen, with a definition for enclosed kitchen. The ACH airflow rate calculation option would no longer be permitted in homes without an enclosed kitchen.

Standard 62.2 includes a compliance pathway for existing buildings in which there is no existing kitchen exhaust. This path is for use when other factors such as cost limits prevent the use of a range hood[7]. This pathway – generally applied in conjunction with energy retrofits – involves increasing the continuous whole-home ventilation rate as a means of making up for the absence of local exhaust.

ASHRAE 62.2 also requires that flows be confirmed through measurement or through a prescriptive pathway. The prescriptive pathway requires that the ducting fall within diameter, length, and material allowances specified in the standard along with a corresponding rated airflow rate of the fan at 62.5 Pascals (Pa) (0.25 inches of water column, or IWC) of static pressure as given by listing in the Home Ventilating Institute (HVI) directory.

Portions of ASHRAE 62.2 have been adopted into California building code, and the US Department of Energy's (DOE) Weatherization Assistance Program (WAP) requires contractors to meet ASHRAE 62.2.

ENERGY STAR Range Hoods

Range hoods can earn an ENERGY STAR rating by moving 2.8 cfm per watt (W) at $\leq$ 2 sone and having a maximum airflow rate of 500 cfm. ENERGY STAR range hoods must be airflow and sone rated using HVI or ANSI/AMCA test procedures[8].

Home Ventilating Institute (HVI)

HVI is a nonprofit trade group that certifies airflows measured using HVI Procedure 916[9] and sound levels measured using HVI Procedure 915[10] and publishes the certified values for a large number of hoods and settings in the HVI Certified Products Directory[11].

HVI also provides guidance on "minimum" and "recommended" airflow rates for wall-backed and island cooktops[12]. For wall-backed range hoods, HVI's recommended kitchen ventilation rate is 100 cfm for every linear foot of range, and their minimum rate is 40 cfm per linear foot of range. For a 30-inch wide cooktop or range, this corresponds to 100 cfm minimum and 250 cfm recommended. For island range hoods, the recommended rate is 150 cfm per linear foot of range, and the minimum rate is 50 cfm per linear foot of range.

International Residential Code (IRC)

The 2012 International Residential Code (IRC)[13] provides specifications, but does not per se require kitchen exhaust ventilation. It states that if local kitchen exhaust is provided, the fan should deliver 100 cfm intermittent or 25 cfm continuous exhaust. The code allows for use of a recirculating range hood as an alternative to exhaust ventilation. The IRC requires that when a range hood or other kitchen exhaust fan exceeds 400 cfm, a make-up air system must be installed.

The IRC has been adopted by multiple US states, in some cases in altered form.

International Mechanical Code (IMC)

The 2009 IMC provides guidance on the installation of range hoods, but does not per se require kitchen ventilation of any kind.

The IMC has been adopted by multiple US states, in some cases in altered form.

International Energy Conservation Code (IECC)

The 2012 IECC[14] states that mechanical ventilation must meet the requirements of the International Residential Code or the International Mechanical Code.

The IECC has been adopted by multiple US states, in some cases in altered form.

International Green Construction Code (IgCC)

The 2012 IgCC states that kitchen ventilation and exhaust systems shall be in accordance with the International Mechanical Code and with additional make-up air requirements for high airflow systems.

Since its inception in 2012, portions of he IgCC have been adopted by several US states and municipalities.

High Performance Homes Programs

DOE Challenge Home

The DOE Challenge Home National Program Requirements (Revision 02)[15] specifies that homes must meet the Environmental Protection Agency's (EPA) Indoor airPLUS Construction Specifications, which in turn require that homes meet ENERGY STAR for Homes or ASHRAE 62.2 requirements for kitchen ventilation.

ENERGY STAR for Homes (ESH)

ENERGY STAR for Homes (Version 3.0, Revision 7) requires that kitchen exhaust is vented to the outside of the home. The ESH kitchen ventilation airflow rates are identical to those of ASHRAE 62.2, with one caveat: unlike ASHRAE 62.2, ESH explicitly defines how to define kitchen volume for calculating the 5 kitchen ACH airflow rate, stating that the "kitchen volume shall be determined by drawing the smallest possible rectangle on the floor plan that encompasses all cabinets, pantries, islands, and peninsulas and multiplying by the average ceiling height for this area." (Energy Star 2013a). This method applies both to enclosed and unenclosed (open floor plan) kitchens.

Like ASHRAE 62.2, ESH allows two compliance pathways: airflow measurement or HVI rated airflow (at 62.5 Pa) alongside the 62.2 prescriptive duct-sizing table. But unlike ASHRAE 62.2, and largely to permit the use of overthe-range (OTR) microwave hoods, ESH currently allows the use of non-HVI-rated fans without requiring airflow measurement. As a provisional measure to facilitate adequate airflow, ESH added a requirement that non-HVI rated kitchen exhaust fans must be vented to outside via a smooth duct having a diameter of at least 6 inches (Energy Star 2013a). ESH publishes a guidance document specifically to help their building partners comply with kitchen ventilation requirements (Energy Star 2013b).

ENERGY STAR for Homes allows for homes that are PHIUS+ certified to use a continuous kitchen fan with an exhaust rate of 25 cfm, citing the 2009 IRC requirements.[16]

Environmental Protection Agency (EPA) Indoor airPLUS

The EPA Indoor airPLUS Construction Specifications[17] (Version 1, Revision 2) (EPA 2013) require that homes meet either ENERGY STAR for Homes or ASHRAE 62.2-2010 requirements for kitchen ventilation.

Leadership in Energy and Environmental Design (LEED) for Homes

The LEED for Homes[18] rating system requires that homes meet ASHRAE 62.2-2007[19] requirements for kitchen ventilation. An extra point is given if the exhaust fan's as-installed flow is measured by a LEED for Homes Green Rater[20].

Passive House Institute US (PHIUS)

Passive House certification is solely performance-based, not prescriptive, and as such has no requirements for kitchen ventilation per se. Based on conversations with Passive House representatives, we understand that the recommended approach to kitchen ventilation in passive houses is to use a recirculating range hood and provide an energy recovery ventilator (ERV) or heat recovery ventilator (HRV) inlet in the kitchen to provide continuous exhaust[21].

Current Gaps in Kitchen Ventilation Codes, Standards, and Ratings

Pollutant Capture Efficiency

The current best practice for dealing with kitchen contaminants is to have and use venting range hoods that efficiently capture the contaminants generated during cooking. Yet, there currently are no requirements, nor even a standard method of test related to the effectiveness residential range hoods at removing pollutants and moisture. Although the literature is not expansive, there are a number of published research papers which demonstrate that capture effectiveness varies widely with the hood morphology, installation location, airflow, and other characteristics (Li and Delsante 1996; Lim and Lee 2008; Madsen et al. 1994; Li et al. 1997; Rim et al. 2012).

Measurements in homes and in laboratory testing indicate that many range hoods have highly varying capture efficiency (CE) for burner exhaust pollutants, depending on the fan speed and whether the front or back burner is used (Delp and Singer 2012; Singer et al. 2012). In particular, it appears that many hoods have substantially higher capture efficiency when cooking occurs on back burners compared with front burners.

Current codes and standards implicitly presume that a kitchen ventilation fan's efficacy at removing cooking pollutants and moisture is largely or wholly determined by its airflow rate. Recent Lawrence Berkeley National Laboratory (LBNL) research on kitchen pollutant capture efficiency (CE)

(Delp and Singer 2012; Singer et al. 2012) suggests that while there is a correlation between a range hood's airflow rate and its CE, the two are not synonymous. Even when inducing the same airflow rate, the CEs among the tested range hood models differed considerably, in some cases by a factor of three. These performance differences are assumed to result from differences in hood morphology (e.g., flat, shallow sump, deep sump) and the extent to which burners are covered.

The study's limited sampling suggests that range hood designs with shapes and geometries that facilitate pollutant capture are capable of achieving 80% capture[22] levels at 200 cfm (double the ASHRAE 62.2 minimum and close to the levels recommended by HVI), even for front burners. The study also showed that for the hood with the most effective capture geometry (deep sump, extending over front burners), the marginal CE increase achieved per cfm increase declines markedly after 250 cfm. This suggests that for the typical residential gas range[23], a well-designed range hood does not need to move more than 250 cfm to be effective. Higher capacity ranges may require higher airflow rates. Further research is needed to determine the appropriate capture efficiency percentage necessary to reduce exposure to cooking pollutants to acceptable levels.

These findings suggest a need for codes, standards, and test protocols that more accurately evaluate a range hood's effectiveness at performing its fundamental function: to remove kitchen pollutants and moisture at a sound level that does not discourage its use. Airflow rate, or "cfm" (cubic feet per minute), has been the de facto measure of range hood performance. A metric that more directly and accurately describes functional performance is needed.

Disparity between Advertised or Rated and Actual, As-Installed Performance

Research in homes (Singer et al. 2012; Fugler 1989; Nagda et al. 1989; Stratton et al. 2012) has found that installed range hoods very often do not move as much air as certified ratings indicate and for advertised airflows that are not certified, the actual installed airflow can be a small fraction of the advertised value (Singer et al. 2011).

We believe that this observed disparity between rated or advertised airflows and measured installed airflows is due primarily to airflow resistance and resulting static pressures in installed duct systems that are significantly higher than those present when the airflow of the fan is rated.

When there is no field-testing to verify the exhaust flow rate, ASHRAE 62.2 requires rated airflows to be tested using the HVI 916 test procedure at

62.5 Pa (0.25 IWC), which is believed to be a reasonable estimate of the static pressure in a typical installed residential ventilation duct system. However, at the time of this writing, only 5 of the 3694 range hoods listed in the HVI ventilation products directory (HVI 2013) have rated airflows for 62.5 Pa of static pressure.[24]

The advertised airflow values in products without certified ratings appear in many cases to be free air delivery, i.e., the airflow generated by the fan without any restriction. Clearly this is not representative of the installed conditions under which the fan will operate. In addition to the ductwork, the hood itself may introduce restrictions that reduce airflow.

Use of Kitchen Exhaust Ventilation in U.S. Homes

There are very limited data about the use of kitchen exhaust ventilation in U.S. homes. The California Energy Commission's research programs have supported survey based studies over the past several years that provide data on kitchen ventilation practices in California (Klug et al. 2011b; Mullen et al. 2013; Offermann 2009; Piazza et al. 2007).

In the Mullen et al. study (2013), among those who said they used their kitchen exhaust system regularly, the top three reasons cited for using exhaust were to remove smoke, odors, and steam/moisture. Among those who reported not using their exhaust system regularly, the reasons cited were that it was not needed, it was too noisy, and that they did not think about it. A minority of respondents reported routinely using kitchen exhaust when cooking. A fraction of people claim to use it "as needed"; but the meaning of "as needed" is subjective and likely varies across respondents.

The Klug et al. study (2011b) asked participants whether or not they used a range hood when a window was not open in the kitchen when cooking breakfast, lunch, and dinner. Forty to 60% of respondents indicated that they did not have a window open and did not use a range hood while cooking.

The Piazza et al. study (2007) surveyed range hood use in new California homes and found that when using the cooktop, 28% of respondents always use the kitchen exhaust fan/range hood, 32% only use it when odor or humidity seems to be an issue, 26% "sometimes" use it, and 13% rarely or never use it. When using the oven, 15% of respondents always use the kitchen exhaust fan/range hood, 12% only use it when odor or humidity seems to be an issue, 15% "sometimes" use it, and 56% rarely or never use it.

The Market Perspective

Achieving the objective of effective management of kitchen contaminants in residences – either through kitchen ventilation or alternative, air-cleaning technologies – is both a market transformation and a technical challenge.

At present, the two largest market players responsible for establishing demand for kitchen ventilation are the public and the bodies responsible for building codes and standards. These groups on the whole appear to have a limited appreciation of the importance of the need to manage kitchen-associated moisture and pollutants for indoor air quality and health protection. For those within each group that recognize that there is a need, established metrics or systematic information about pollutant and moisture removal effectiveness of products and system designs are lacking.

How the public and code-setting bodies conceive of kitchen exhaust ventilation is not entirely clear. Surveys indicate that the most common reasons for using range hoods are to remove moisture, smoke and odors (Klug et al. 2011b; Mullen et al. 2013). We infer from these results that many people assume that range hoods should serve those functions, even if they do not consider how or if the functionality is achieved. Existing standards recognize the importance of airflow as a design element that impacts effectiveness and also recognize that operational loudness is an important characteristic that may limit the use of kitchen exhaust fans. There is evidence that some fraction of the public recognizes a connection between airflow and effectiveness and associates higher airflows with better performance. The connection is communicated to purchasers of high-output cooking devices who are advised to purchase range hoods with higher airflows, and may be inferred by consumers shopping for higher end ventilation products that prominently advertise high airflow rates. Yet we infer that most consumers and code-setting bodies do not see kitchen exhaust ventilation as an essential building service because (a) so many states lack code requirements and as a result, so many new homes are constructed without any kitchen exhaust ventilation; (b) the equipment that is most commonly installed is likely not very effective as installed; and (c) many people do not routinely use kitchen ventilation even when it is available in their homes. In the absence of code requirements, and with homebuyers in many areas not recognizing kitchen exhaust ventilation as an essential building service, the home building and renovating industry has little incentive to incur (and thus have to recover) the costs of providing more effective kitchen exhaust ventilation systems.

The most broad-based demand signal in the area of kitchen ventilation appears to be the desire of homebuyers for over-the-range (OTR) microwaves. This demand signal is inferred from the large fraction of new and recently built homes that feature these appliances (Klug et al. 2011a) and from the anecdotal observations from our webinar participants. OTR microwaves have an internal fan that can be activated to draw air from below, up through grease screens, then directed either to an exhaust duct or recirculated back into the kitchen. When set up in exhaust mode, at least some of the OTR microwave range hoods pull air from another location in addition to pulling air through the bottom of the unit. Though systematic data is lacking, our hypothesis from talking with building professionals and homeowners around the US is that the OTR microwaves are most commonly installed to operate as recirculating range hoods, as opposed to exhaust devices. Our understanding is that these appliances are commonly shipped with the internal airflow baffles set to recirculation mode and we have anecdotal evidence that some OTRs that are installed with venting to the outdoors may nevertheless be operating as recirculation devices because the internal baffle was not adjusted to direct airflow to the vent. We are unaware of any research quantifying their effectiveness in reducing cooking-related smoke or odors; but anecdotal experience suggests that they can have at least some impact on visible smoke. Capture is impeded by the basic design premise of an OTR: flat bottom, small intakes and filters, and insufficient depth to extend beyond the front burners.

In reviewing web-based marketing materials – including manufacturer and retailer web sites and product reviews – we found little attention paid to the performance of the OTR microwaves as range hoods. It is relatively common for the product specifications to list only a single airflow (at the highest setting) with no information provided about loudness. These advertised airflows are not HVI-rated and at least some products appear to advertise the fan's free air delivery or an airflow rate at no duct static pressure, which is not indicative of the measured airflow of the device as installed.

Without a market impetus for devices that achieve high or even moderate performance for pollutant and moisture capture, innovation in this area is challenged.

Pathways to create a robust demand for effective kitchen exhaust ventilation products include (1) establishing codes and standards with appropriate performance-based requirements and (2) raising awareness of the importance of kitchen exhaust ventilation and specifically of the performance characteristics that impact effectiveness for indoor air quality protection. The public education required to achieve the second pathway should also help with the first.

Innovation in kitchen ventilation appears to be focused at least in large part on range hoods that are suitable complements to higher-end cooking devices. These cooking devices range from residential models with large burners (e.g., a total cooktop capacity of 50,000 to 60,000 British thermal units per hour (Btu/h) as compared to a base model 30" cooktop with four 9,000 Btu/h burners) to commercial style cooktops and ranges that can have 6 or more cooktop burners with total capacity in excess of 60,000 Btu/h.

RECOMMENDATIONS

The following section begins with our recommendations for key Objectives to bring about a transition to reduced kitchen pollutant exposure in homes. Each Objective is supported by one or more Actions, Research topics, and Technologies. The first major section comprises Objectives; Actions; and Research, Development, and Deployment needed to transform the kitchen ventilation market. The second major section identifies Technology Developments needed to improve kitchen ventilation systems in homes and to ensure their effectiveness.

Objectives and Needs for Market Transformation

01. Expand Awareness of the Importance of Kitchen Ventilation

A foundational need for improving the availability, use and performance of kitchen ventilation in U.S. homes is widespread awareness that it is necessary and beneficial. There is a need for recognition both by the public and among building industry professionals – including builders, architects, contractors, and realtors – that natural gas cooking burners, electric coils, and cooking of food produce pollutants that can reach hazardous levels when effective kitchen ventilation is not available and used in homes. Achieving this recognition will spur the advances described later in this document.

There are many stakeholders that can contribute to this objective, including but not limited to the following:

- Home performance industry associations (including Affordable Comfort, Inc. (ACI), the Building Performance Institute (BPI), and the Residential Energy Services Network (RESNET));
- Builders and building industry associations; contractors and contractor associations;
- Public health agencies (e.g., state and local departments of public health, state and federal environmental protection agencies);
- Public housing agencies;
- Non-governmental organizations working to promote healthy homes;
- Industry associations (e.g., HVI, ASHRAE);
- Manufacturers, distributors, and retailers of ventilation equipment;
- Industry associations, manufacturers and distributors of gas and electric cooking appliances.

Tracking progress on this objective is not easy because there is no established mechanism to collect data on individuals' understanding of indoor air quality hazards or their practices that impact indoor air quality (IAQ). This is not to say that there are not well-established survey methods that could be applied to this objective; but that it has not been done regularly in the past and that establishing the content and structure of a periodic survey on the public's understanding of IAQ would be a significant undertaking. This is an important research need. Related: A1, A2, A4, A7, R2, R4, R5, R7, R8, T1, T5

02. Expand Use of Kitchen Ventilation

As more people become aware of the need for kitchen ventilation, the presumption is that they will more often use whatever ventilation is available to them. Use of currently-installed systems and devices that are not effective at quickly and effectively removing odors, smoke, etc. (i.e., the indicators of pollutant generation that can be detected without instrumentation) and/or that are deemed to be too loud may help create a demand for better products and information about product performance. Efforts to drive demand through public education efforts need to be coupled with increased availability of meaningful performance metrics and affordable, effective range hoods.

Efforts to expand use of kitchen ventilation logically should be connected to efforts to raise awareness of the hazard and to the distribution of information about selecting and using kitchen ventilation elements. The list of stakeholders for this objective is thus the same as for the first objective described above. Similarly, tracking progress on this objective will not be easy because there is no established mechanism to collect data on individuals' understanding of indoor air quality hazards or their practices that impact IAQ. This is an important research need.

Related: A1, A3, A6, A7, R2, R3, R4, R5, R6, R8, T1, T2, T3, T6, T7

03. Publicize Existing Guidance for Building Professionals on the Basics of Kitchen Ventilation and Develop Guidance on Best Practice Designs and Equipment Specifications

The appendix contains a summary of available relevant codes, standards and guidance documents, many of which are discussed below.

The Building America Solution Center (BASC) has very good basic guidance on kitchen exhaust ventilation that references the IRC, ASHRAE 62.2, and ENERGY STAR. This represents something short of best practice but certainly exceeds the current minimum standard.

ASHRAE 62.2 is a consensus standard that defines minimum requirements for kitchen ventilation for the objective of achieving acceptable indoor air quality. Among those who recognize the need for kitchen ventilation, there is a growing interest and call for guidance on best practices. There are questions about the relative performance of general kitchen exhaust ventilation as compared to range hoods, about the effectiveness of recirculating hoods that include activated carbon or other features designed to remove pollutants, and about how to select range hoods that maximize pollutant capture efficiency.

Information and analysis is available to enable development of best practice guidance. We know that range hoods are potentially much more effective than general kitchen exhaust at capturing pollutants and moisture before they mix into the home and elevate exposure and health risk. For product selection, building professionals can refer to the same resources that will be made available to consumers. In consideration of varying performance, builders and designers could offer more options and/or select standard equipment that is suitable to the cooking appliance. For example, in homes with counter cooktops, a standard depth range hood may be adequate. In homes with slide-in ranges a deeper hood is warranted. More attention should be paid to the installation of low-pressure drop ducting, including transitions and roof and wall caps.

Research is needed to develop guidance on the use of recirculating hoods. Simulations conducted to explore the relative impacts of general kitchen vs. range hood ventilation need to be analyzed and translated into guidance. Guidance should be provided not only for the ideal case of a new home, but also for retrofitting existing homes, including challenging situations in which there is not currently any kitchen exhaust and the installation of kitchen exhaust may be problematic.

Related: A2, A4, A5, A7, R1, R3, R4, R5, R7, R8, T1, T3, T4, T5, T6, T7

04. Provide Better Performance Information and Guidance to Purchasers and Users of Kitchen Ventilation Products

As the public and building industry professionals become aware of the need for kitchen exhaust ventilation, it is important that they have access to accurate and clear guidance – supported by accurate and reliable data – on how to select and operate effective kitchen exhaust ventilation.

Available research results and analysis can be used to develop guidance about selection and use of hoods to maximize performance. Research indicates that many hoods have high capture efficiency on back burners but performance on front burners is more variable across hoods and settings. Research also indicates that achieving capture efficiencies of 80-90% or higher – even on front burners – requires roughly 200 cfm of airflow (substantially more than the 62.2 minimum air flow rates). Further research is needed to determine the appropriate capture efficiency percentage necessary to reduce exposure to cooking pollutants to acceptable levels. We know that loudness is among the top reasons that people do not use the equipment. There is solid experimental evidence that performance for front burners improves greatly when the range hood covers the entirety of the front burners.

Related: A2, A3, A4, A5, A6, R2, R3, R4, R5, R7, R8, T1, T2, T3, T4, T6, T7

05. Improve and Expand Accessibility of Guidance for Installers and Inspectors

Problems with performance in many systems could be linked to improper installation. There is anecdotal evidence that duct static pressure in many home venting systems greatly exceeds the 25 Pa (0.10 IWC) under which hoods are tested for airflow certification. These high pressure drops can result from long and/or convoluted duct runs, high pressure drop transitions and roof/wall caps and even from backdraft dampers that are inhibited from opening fully. Building inspectors can play a crucial role by helping to ensure that hoods and vent systems are installed correctly.

As part of installation guidance, the importance of commissioning should be stressed. As noted elsewhere, research and technology developments are needed to improve commissioning tools and practices to enable widespread application.

ASHRAE 62.2 requires measurement of flows or compliance with prescriptive limits on duct pressure drop as indicated by equivalent length. The standard includes tables that estimate the actual airflow as a function of rated airflow at 62.5 Pa (0.25 IWC) and equivalent length. As noted elsewhere, the current problem with this approach is that very few range hoods are HVI rated at 62.5 Pa.

Related: A1, A2, A4, A5, R1, R3, R4, R5, R7, R8, T1, T3, T4, T5, T6

06. Raise Awareness of Kitchen Ventilation Requirements of ASHRAE Standard 62.2

Much of the focus of ASHRAE 62.2 over the years has been on the requirements for residence level air exchange rates. Anecdotal reports (e.g., from session discussions and conversations at home performance conferences) suggest that many practitioners are only vaguely aware about the requirements related to kitchen ventilation. Increasing awareness of these requirements should help to expand the installation of kitchen exhaust ventilation in US homes.

ASHRAE 62.2 can be met using an exhaust fan in the kitchen that moves 5 kitchen air changes per hour. The exhaust fan can be continuous or on-demand (intermittent). The standard is currently ambiguous about how to determine the kitchen volume in homes without enclosed kitchens (i.e., open floor plan homes).

A change proposal for the kitchen area exhaust ventilation requirement has been submitted to the 62.2 committee with the intent of simplifying this pathway for improved compliance and exposure reduction. The proposed change would require 300 cfm of kitchen exhaust if the fan is not a range hood, 100 cfm of kitchen exhaust if it is a range hood, or 5 ACH of kitchen exhaust based on kitchen volume for an enclosed kitchen, with a definition for enclosed kitchen. The ACH airflow rate calculation option would no longer be permitted in homes without an enclosed kitchen.

The standard currently includes a provision for existing homes without kitchen exhaust to make up for the deficit by a proportionate increase of whole-home ventilation. While clarifying these requirements will advance efforts to have more homes comply with the current version of this minimum standard, there is a need to substantially rework the standard to be based on capture efficiency, rather than airflow. The standard also should reflect that increasing continuous ventilation rates may not substantially mitigate acute exposures. An improvement over the current retrofit allowance may be to require a controller in the kitchen if a higher airflow exhaust fan is used at another location in the home.

California is among a handful of states with building codes that require new homes to meet ASHRAE 62.2 kitchen ventilation requirements. In practice, however, it seems unlikely that new California homes are strictly meeting the kitchen ventilation requirements. To use the ASHRAE 62.2 prescriptive compliance path, range hoods must be HVI rated at 62.5 Pa (0.25 IWC). To our knowledge, there are currently no HVI-rated microwave range hood models available and only 0.13% of HVI-certified range hoods are rated at this duct static pressure. Therefore the airflow of nearly every range hood installed in California must be measured as installed in order to meet California's Building Energy Efficiency Standards: Title 24, Part 6[25]. For a variety of reasons, accurately measuring the installed airflow of a range hood is notoriously difficult (Stratton et al. 2012), making it unlikely that many range hood airflows are being measured.

Related: A2, A5, R1, R4

07. Establish Capture Efficiency As a Key Performance Measure for Range Hoods

The metric used to define performance of kitchen ventilation should more directly correspond to the purpose of the ventilation: removal of moisture and contaminants. The most intuitively simple and direct metric is first pass capture efficiency (CE) for range hoods. This metric describes the fraction of contaminants generated at the cooktop (also applicable to the oven for a slide-in range) that are removed by a range hood before mixing into the bulk kitchen air volume.

Research has shown that range hoods operating at current ASHRAE 62.2 minimum air flow rates can have very low capture efficiency (e.g., 30% or less) when cooking occurs on front burners (Delp and Singer 2012; Singer et al. 2012). Higher airflows generally improve capture efficiency; but the relationship varies with cooking location and range hood design. The most direct approach to ensuring that range hoods accomplish their prime performance objective is to measure that performance directly. A recognized standard test method will enable the specification of performance requirements in minimum standards such as ASHRAE 62.2 and also in programs that seek to specify exceptional performance. A test method for capture efficiency could be used by range hood manufacturers to design, develop, and advertise products. HVI-rated capture performance could be prominently touted in product literature. In the longer term, capture efficiency may become just one component of a more comprehensive contaminant removal metric for cooking and burner exhaust pollutants. Such a metric would consider not just first-pass pollutant capture, but the removal of kitchen pollutants from the home over a given time period. This metric would allow for direct efficacy comparisons among varying kitchen ventilation strategies, including local wall or ceiling ventilation, recirculating range hoods with pollutant removal, and whole-home ventilation, in addition to vented range hoods.

Related: A3, A4, R3, T7

08. Promote Development of Automatic Operation As a Future Standard Requirement

There was broad agreement among meeting participants that the automation of kitchen ventilation would enable more robust indoor air quality protection. There was much less agreement about specific interim objectives and the urgency of pursuing standards related to automation. Already on the market are several range hood models with an automatic operation algorithm and more models are anticipated. Automated operation has the potential to help ensure reduced exposure to kitchen pollutants, but the technology must be paired with range hoods that are quiet and have geometries that allow effective capture efficiency at lower flows. Automated range hoods that are noisy have the potential to incite public backlash, especially if they are mandatory.

LBNL has conducted preliminary laboratory evaluations of some range hoods with automatic operation functionality; these tests indicate variable performance across the available models with some technologies showing promise. To ensure efficacy and consistency among automated range hoods, a standard test method should be developed.

Related: A6, R3, R6, T6, T7

09. Develop Tools and Guidance to Expand Commissioning of Kitchen Ventilation Systems

In order to better evaluate how kitchen ventilation systems are performing as installed, we need to be able to commission them quickly, accurately, and reliably. The primary metric that is measured is installed airflow of range hoods. In addition, technicians should be able to reliably and accurately measure sound level and duct pressure. Duct systems with high static pressure waste fan energy and lead to poor performance. Pressure measurement can be used as a diagnostic indicator: if the pressure is much higher than expected, it can indicate to the technician that flow is being impeded.

Related: A1, A7, R1, T5

O10. Reduce the Number of New Homes Built without Kitchen Exhaust Ventilation

The most direct way to do this would be to change code, starting with the uniform mechanical code. Increased awareness of the hazard, by building professionals and by the public, will help motivate action. The overarching goal is relatively simple: make sure that the vast majority of new homes have at least some minimal amount of kitchen exhaust ventilation with low-pressure drop ducting that can accommodate hoods with airflows exceeding 200 cfm. The recommended minimum requirement is for an ASHRAE 62.2 compliant range hood with low-pressure drop ductwork verified by inspection, and performance verified by commissioning. This will require a sea change in practice in areas of the country that currently do not require or commonly incorporate kitchen exhaust ventilation in new homes.

For maximum impact, effort should be focused on states that currently do not have kitchen ventilation requirements in their codes and that are projected to build many new homes in the coming years. A cursory review of state building energy codes suggests that currently fewer than 25% of US residents live in states whose building codes require kitchen ventilation in new homes. As we work to increase the number of states with building codes requiring kitchen ventilation, we also need to provide code inspectors with the knowledge and tools to verify airflow, sound and make-up air requirements.

Related: A1, R1, R3, R8, T1, T2, T3, T6, T7

O11. Incorporate Kitchen Ventilation As a Standard Element of Energy Retrofits

At least one retrofit program - DOE's Weatherization Assistance Program (WAP) -- requires ASHRAE 62.2. Kitchen ventilation is considered an allowable expense for WAP projects.

The Building Performance Institute (BPI) Building Analyst Standard (BPI 2012) requires that work scopes for homes that have unvented gas ovens in enclosed spaces without mechanical ventilation include a recommendation for the addition of exhaust ventilation. The BPI requirements have historically been primarily concerned with exposure to CO, and have not discussed any health concerns related to other cooking pollutants. The standard contains no ventilation recommendation for kitchens with electric ranges. The next BPI standard will include references to ASHRAE 62.2-2013, including requirements for local kitchen ventilation.

The Residential Energy Services Network (RESNET) has no kitchen ventilation requirement but a new version of the RESNET standard soon to be released will require that ASHRAE 62.2 be met; this will include the kitchen exhaust ventilation requirements. Leading home performance organizations such as BPI, RESNET, and Affordable Comfort, Inc. (ACI) are essential partners in teaching home performance practitioners about the importance of effective kitchen ventilation. These and other industry organizations will be key players in raising awareness of the need and opportunity to include kitchen ventilation as an IAQ measure during retrofits.

Since many homeowners are more concerned with IAQ and health than with energy, there also may be business opportunities for contractors to expand scopes of work for energy retrofits by including kitchen ventilation as an IAQ measure. A retrofit is a good time to add kitchen ventilation for several reasons. First is that air sealing to reduce infiltration will also result in higher time-averaged concentrations of pollutants generated in the home. Second is that the tradespeople needed to install a kitchen ventilation system may already be coming to the home; this translates to lower cost. Third is that the energy retrofit contractor will already be looking to seal any existing penetrations associated with kitchen ventilation or appropriate air sealing can be applied to the new installation. Fourth is that the energy retrofit contractor will already be conducting any combustion safety checks that should be conducted when adding or replacing a large exhaust fan. One issue affecting these retrofits will be the potential to cause depressurization with higher-airflow range hoods. Tight envelopes and powerful range hoods are likely to cause failures of worst-case depressurization tests. Effective low-flow range hoods, dedicated make-up air, and automatic turn-off are all potential solutions for this problem. A related issue is to ensure that dedicated, automated make-up air – e.g., via pressure relief damper, or inter-locked supply fan – is always installed with high airflow (>400 cfm) range hoods, as required by manufacturers' instructions and building and mechanical codes[26]. However, research has found that such high flow rates are not necessary to achieve adequate levels of capture efficiency (Delp and Singer 2012).

Related: A7, R2, R4, R5, R8, T2, T3

012. Develop and Demonstrate Kitchen Ventilation Systems for Low- and Zero-Net-Energy, High Performance Homes

The envelope tightness and desire for low energy consumption in high performance new homes and deep energy retrofits require that particular attention be paid to these homes' kitchen ventilation systems. In these high performance homes, kitchen ventilation must not interfere with energy savings and must not cause indoor air quality and health problems. This is the "first do no harm" mandate. Depressurization of a home can occur when a kitchen exhaust fan pulls indoor air out faster than the building envelope can let outdoor air in. Backdrafting of natural draft appliances is the primary concern with depressurization, but because of their low efficiency, natural draft appliances are not usually found in high performance homes. Excessive depressurization can be a problem even in homes that do not have combustion appliances; it can cause uneven space conditioning, increase space conditioning energy demand, and can pull pollutants from attics, crawlspaces, and interstitial cavities into the living area of the home. Avoiding depressurization can be accomplished through use of a pressure relief damper, by providing make-up air, or by avoiding very high airflow range hoods. An automatic / timed shut off feature would limit the hazard associated with depressurization by eliminating long durations of high exhaust flows resulting from a range hood being left on inadvertently. As part of a comprehensive effort to reduce energy consumption, many high performance homes are seeking to reduce the need for kitchen ventilation by reducing pollutant sources. One method being employed to reduce cooking pollutants is the use of electric induction instead of electric resistance or natural gas cooktops. Recent measurements (Less 2012) suggest that induction cooktops produce fewer particles than natural gas or electric resistance cooktops; however, the cooking activities still produce moisture and pollutants that need to be removed from the living space (Fortmann et al. 2001). Also promising – though yet unproven – is the concept of a recirculating range hood with effective air pollutant removal technology. The development and adoption of effective, low-flow venting hoods or recirculating range hoods with air cleaning would enable the protection of IAQ without adversely impacting energy consumption in high performance homes. Moisture would still need to be resolved through dehumidification and/or whole-home ventilation.

Related: A1, A2, A3, A4, A5, A6, R3, R6, R7, R8, T1, T4, T5, T6, T7

Specific Actions and Targets for Market Transformation

A1. Make Kitchen Ventilation Required in International Residential Code (IRC), International Mechanical Code (IMC), International Energy Conservation Code (IECC), International Green Construction Code (IgCC), and Uniform Mechanical Code (UMC)

The IRC, IMC, IECC, and the new IgCC are at the core of building codes for the majority of US states. The UMC is used by many local jurisdictions.

Currently these building codes do not require kitchen exhaust ventilation. Thus, the default practice in most of the US is to not require kitchen ventilation. If these codes were altered to require kitchen ventilation, then that would become the new "norm" for building codes in the US.

A2. Increase Awareness of Airflow Rating Certification; Ensure That Best Practice Guides and Home Performance Standards Specify Certified Airflows at Appropriate Duct Pressures

Members of the public are likely to be unaware of the difference between an advertised airflow and an airflow that has been rated by an independent third party, such as HVI. Even most home performance professionals may not know to distinguish between an HVI-rated airflow at 25 Pa (0.10 IWC) and one rated at 62.5 Pa (0.25 IWC).

One method of improving understanding of effective kitchen ventilation within both audiences would be the development of best practice guides

Guidance on current best practice kitchen ventilation might include the following, based on the findings of Singer et al. (2012):

Hood Characteristics

- Capable of 200 cfm flow, as rated by HVI at 62.5 Pa (0.25 IWC)
- Quiet operation (<3 sone; lower preferred) at 200 cfm or more
- Hood covers front burners
- Geometry that facilitates plume capture

Behaviors

- Use kitchen ventilation during and after every cooking event
- Cook on back burners to assure/increase removal of cooking emissions
- Clean grease screens regularly to maintain flow

This guidance could also be added to HVI's Consumer Guide for Range Hoods[27].

An easily searchable directory of kitchen ventilation products would enable consumers and homebuilders to make more informed decisions when choosing a range hood or kitchen exhaust fan. HVI currently publishes such a directory, but its current format[28] may be confusing, too long, and difficult for lay audiences. Ideally, this database would be made easily searchable online, with filters for categories such as manufacturer, device type, rated sone level, and rated airflow at 62.5 Pa.

A3. Facilitate Development and Adoption of Standard Method of Test for Capture Efficiency (Pollutant Removal Effectiveness)

In order for capture efficiency to be used as a performance metric, a consensus capture efficiency standard method of test must first be developed. A 2012 study by LBNL (Delp and Singer 2012) discusses a capture efficiency test methodology which could be refined into an ASHRAE/ASTM standard method of test. HVI could certify these capture efficiency tests and results. Range hoods would then be labeled with an HVI capture efficiency rating through industry-wide collaboration and support of standard bodies and code authorities.

A4. Incorporate Capture Efficiency into Standards Including Ashrae 62.2, Energy Star for Range Hood Products, and High Performance Home Standards

Once capture efficiency is established as a performance metric for range hoods, required and recommended CE performance levels can be cited in relevant best practice guidance and standards, including HVI recommendations, ASHRAE 62.2, ENERGY STAR for Homes, EPA Indoor airPLUS, and the International Residential Code.

HVI ratings would cite a product's capture efficiency and sone level, in addition to rated airflow. These performance ratings should be displayed prominently on clear, uniform labels on range hood products to that people can easily compare the performance of different models while shopping.

A5. Align Home Ventilating Institute (HVI) Airflow and Sound Test Procedures and Ashrae Standard 62.2 Prescriptive Compliance Pathway to Achieve Effective Performance of Installed Range Hoods

HVI test procedures, ASHRAE 62.2 requirements, and actual installed performance of range hoods need to become aligned in order to improve kitchen ventilation performance in homes. The current disparities between these three factors are impeding progress toward the overarching goal of reducing people's exposure to cooking pollutants.

Currently, many range hood models only have an HVI-rated airflow at a static pressure of 25 Pa (0.10 IWC)[29]. Rating an HVI-certified device for airflow at 62.5 Pa (0.25 IWC) of static pressure is optional. However, the ASHRAE 62.2 prescriptive duct requirements are based on an HVI-rated airflow at 62.5 Pa of static pressure[30]. This higher number is likely more representative of the static pressure found in most kitchen ventilation duct systems, and should therefore be used to estimate installed airflow.

Range hoods that are only rated for airflow at 25 Pa (0.10 IWC), cannot currently be evaluated for ASHRAE 62.2 compliance using the prescriptive duct requirements, and must instead be evaluated for compliance by measuring their airflow rate as installed.

There are several options for enabling ASHRAE 62.2 compliance of range hood models whose airflow is currently only rated for 25 Pa (0.10 IWC). One option is for HVI to change the mandatory rating point to 62.5 Pa (0.25 IWC). This would require manufacturers to have 62.5 Pa (0.25 IWC) airflow ratings performed by HVI on many existing models (and is therefore considered by some to be a non-starter).

Another potential option could be to modify the ASHRAE 62.2 standard to add 25 Pa (0.10 IWC) rated flow figures to the prescriptive duct requirements table. These 25 Pa (0.10 IWC) figures would need to be calibrated to ensure that the devices would meet the current airflow requirements when the duct static pressure is 0.25 IWC.

This latter option of adding 25 Pa (0.10 IWC) ratings to ASHRAE 62.2 would be preferred by manufacturers because it obviates the need and expense of additional testing, but this option perpetuates a test condition – 25 Pa (0.10 IWC) of static pressure – that we do not think is reasonable and should be phased out. A potential hybrid approach could entail the addition of the 25 Pa (0.10 IWC) ratings in ASHRAE 62.2 for use with existing product models, and a change to a mandatory 62.5 Pa (0.25 IWC) HVI airflow test requirement, to be used for all future airflow rating of range hoods.

A6. Incorporate Automatic Operation into Best Practice Guides and High Performance Home Standards As Available Products Are Determined to Be Reliable Based on Standard Test Method

Once a standard method of test is established and automated range hoods become widely available on the market, kitchen ventilation best practice guides (including HVI's guide) should incorporate recommendations for the use of automated devices. High performance home standards, such as LEED for Homes, ENERGY STAR for Homes, DOE Challenge Home, EPA Indoor airPLUS, and Passive House should incentivize or require inclusion of automated kitchen ventilation.

A7. Modify Residential Combustion Safety Test Protocols to Reflect Our Current Best Understanding of Pollutant Health and Safety Risks

The current combustion safety diagnostic protocols from the BPI and RESNET primarily consider backdrafting of combustion appliances caused by depressurization. A 2012 review of combustion safety literature (Rapp et al. 2012) concluded that widely-used combustion test protocols tend to overestimate the number of homes whose vented combustion appliances pose a safety risk. The worst-case depressurization test used in these protocols may not represent an accurate determination of risk. The test entails simultaneously operating every exhaust device in the home and observing depressurization and backdrafting of vented combustion appliances. The risk determination does not consider the frequency of this worst-case condition, which is likely rare. It does not consider that the depressurization pulling combustion gasses into the house is also simultaneously diluting those gases with outdoor air.

There is concern that these tests have the potential to discourage people from installing adequate kitchen ventilation, for fear that the vented range hood will cause depressurization-induced backdrafting of the vented combustion appliance(s) and expose occupants to combustion pollutants.

BPI/RESNET combustion safety test protocols should be modified to better reflect our current best understanding of pollutant health and safety risks in homes. In many cases, because it is unvented, the gas range is the home combustion appliance that people should be the most concerned about. Among the modifications to the combustion safety test protocols should be a consideration of whether or not the range hood has a timed automatic shut-off feature. When homes are being built or modified, to obviate the concerns about combustion appliance backdrafting, the placement of atmospherically vented water heaters in areas within or adjacent to the living space should be discouraged or prohibited. Both ASHRAE 62.2 and the IRC require the installation of a make-up air inlet in homes that have range hoods with rated flows of 400 cfm or greater.

Research, Development, and Deployment (RDD) to Support Market Transformation

R1. Develop and Demonstrate Effective Methods and Devices for Commissioning Range Hoods

To ensure that kitchen ventilation systems are performing as designed, it is a high priority to develop easy-to-use tools and methods for measuring range hood flows. ASHRAE 62.2 requires commissioning of installed ventilation systems. Starting with the 2010 version of the standard, ASHRAE 62.2 required that whole-home ventilation airflows be measured. Kitchen ventilation systems can be evaluated via the prescriptive duct-sizing pathway, assuming that the range hood is HVI-rated at 62.5 Pa (0.25 IWC), but since so few range hoods are rated at 62.5 Pa, nearly all range hood flows must be measured as installed to meet ASHRAE 62.2 local kitchen ventilation requirements. Anecdotal reports are that commissioning rarely happens, in large part because of a lack of methods and tools.

The fact that range hood dimensions and configurations vary makes it difficult to measure their inlet airflows accurately. Roof or wall cap dimensions are more uniform, but the outlet airflow measurement approach presents its own safety and measurement accuracy issues[31].

Given that range hood dimensions are likely to continue to be non-uniform, the home performance industry needs a flow capture hood that is versatile enough to effectively capture and measure the inlet flow of a wide variety of range hood designs. The product should also be affordable (i.e., less than $1000), accurate (within ±10% of flow measurement), and professional in appearance[32].

Given that many range hood airflows cannot currently be measured easily, we should make it easier to properly use the ASHRAE 62.2 prescriptive pathway for local kitchen exhaust. Anecdotes suggest that building inspectors are using the HVI-rated flows listed on the range hoods to evaluate compliance with ASHRAE 62.2, even though in almost every case those flows are rated at 25 Pa and the standard requires flows to be rated at 62.5 Pa. Adding to the ASHRAE 62.2 prescriptive duct-sizing table required flows rated at 25 Pa would call to attention the distinction between the two static pressures, thus facilitating proper fan sizing and increasing compliance rates. Energy Star for Homes also has a compliance pathway involving kitchen ventilation commissioning (Energy Star 2013a) that would benefit from improved commissioning tools and methods.

The addition of on-board diagnostics to kitchen ventilation systems has the potential to ensure performance not just at installation but also over time.

R2. Characterize in-Field Use and Performance of Range Hoods

We currently have a poor understanding of the kitchen ventilation systems in U.S. homes. To inform policy development and track progress, we need improved data collection methods and databases on the availability, use, and performance characteristics of kitchen ventilation. A starting point would be to collect baseline data on kitchen ventilation currently installed in US housing stock, including ventilation type and frequency of use. Ideally this data should be available for each state, as the practices depend on local conventions and building codes. At the outset, a mechanism to update data regularly should be included to track changes over time.

A survey of building occupants is one possible approach to gathering data about residential kitchen ventilation systems. A series of questions about the presence and use of kitchen ventilation could be incorporated to the Residential Energy Consumption Survey (RECS). Or a large-scale kitchen ventilation-specific survey could be independently developed and carried out.

Understanding current practice for kitchen ventilation in very high performance homes (including Passive House) is of particular interest.

R3. Develop a Comprehensive Contaminant Removal Effectiveness Metric

Currently, there is not an easy way to empirically evaluate the effectiveness of different kitchen ventilation strategies and systems. There may be a need to develop a metric that is more broadly applicable to kitchen ventilation whether it be in the form of a range hood, a kitchen area exhaust fan, or potentially a filter or other air cleaning device. A more generic contaminant removal effectiveness (CRE) metric would allow more direct comparisons between different design approaches to meeting the need for pollutant and moisture removal from kitchens. It would also enable the setting of more robustly equivalent standards for these various compliance pathways.

One challenge with this approach is that effectiveness will vary for moisture and various pollutants. For moisture, there is substantial buffering capacity in the walls, cabinetry and other materials in the kitchen and other areas of the home. Any moisture that is not removed immediately at the source may in some cases be "captured" by materials on time scales that are competitive with area ventilation. This is not necessarily helpful as this process of moisture capture by materials creates opportunities for mold growth. Some pollutants also are removed from the air through interactions with material surfaces on time scales that could compete with removal via increased area ventilation or filtration. Key considerations for a CRE approach would be (1) to define the frame over which removal is quantified and (2) to specify the contaminants to which the CRE applies. The time frame could be related to periods over which harm can be realized, with harm defined, e.g., as acute pollutant exposures or the time scale for water vapor to bind to residential materials.

R4. Improve or Validate Ashrae 62.2 "Deficit Make-up" Approach for Existing Buildings

In existing homes, ASHRAE 62.2 (ASHRAE 2010) currently allows homes without local kitchen exhaust to meet the standard by increasing the whole-home ventilation by an amount equal to one quarter of the local exhaust rate usually required. An evaluation of the efficacy of this approach needs to be undertaken. There are reasons to suspect that this may not be a good approach, for both IAQ and energy conservation. For example, if a home with an enclosed kitchen requires a 100 cfm local exhaust fan, one quarter that rate – 25 cfm -- would then be added to the total whole-home continuous ventilation rate required by the standard. If the whole-home continuous ventilation rate is 60 cfm, then the total continuous ventilation rate would need to be 85 cfm to meet the standard.[33] This approach is problematic for two primary reasons. First, given the same airflow, general whole-home ventilation is not as effective at removing cooking pollutants as local ventilation at the pollutant source, and the cook and kitchen occupants would be exposed to elevated levels of pollutants before they are removed. Second, ventilating continuously at 85 cfm would substantially increase the space conditioning energy requirement without substantially improving IAQ.

This evaluation may entail a monitoring or simulation study to assess effectiveness of this "deficit make-up" approach to kitchen ventilation.

R5. Evaluate Combustion Safety Allowance for Timed Auto-off Kitchen Ventilation

Currently, the BPI and RESNET combustion safety diagnostic protocols have restrictive allowances for home depressurization to prevent backdrafting of vented combustion appliances. These restrictions may discourage or prohibit the installation of effective kitchen ventilation. A timed "auto-off" feature for range hoods would reduce the risk of sustained depressurization-induced backdrafting in homes; the utility of this feature should be considered in the BPI and RESNET combustion safety protocols.

To evaluate whether such a modification would be justified, a simulation-based and/or monitoring study could be carried out to assess the effect of an auto-off feature on pollutant exposures.

R6. Develop Standard Test Method and Performance Metrics for Automatic Operation

Range hoods with automatic operation have the potential to significantly decrease exposure to kitchen pollutants in homes. However, there is currently no consensus on what it means for a range hood to be automated, and no established metrics for evaluating the efficacy of an automated range hood.

Criteria for minimally effective automatic operation could be developed by considering pollutant and moisture emission rates, and the impacts of various time delays before operation is automatically initiated. Credit should also be provided for hoods that extend operation for some period after a cooking sensor has detected an end to cooking. The findings of this study would then be used as the basis for the development of a standard method of test and performance metrics for automated range hoods. These performance metrics should consider capture efficiency and energy consumption in terms of both conditioned air and fan energy, and models that achieve a designated CE while using the least energy should be prioritized. This test method could become an ASTM/ANSI standard and certified by HVI or another third-body organization.

A study could then be performed to evaluate the performance of the few automated range hoods that are currently commercially available. In addition, range hoods with novel automation mechanisms could be developed and evaluated alongside the commercial products. Long-term performance and health impacts could be evaluated through modeling.

This process could be used as a forum for demonstration of automated kitchen ventilation systems and technologies.

R7. Re-Evaluate Kitchen Ventilation and Moisture Removal

The removal of cooking-generated moisture has always been considered one of the primary functions of kitchen ventilation. If recirculating range hoods capable of effectively filtering particles and pollutants of concern can be developed, then the question remains: what about the moisture? Does it still need to be removed? The answer to that question has long been presumed to be 'yes'.

However, it is not clear that cooking-generated moisture must be removed in all homes in all climates. The capacity for dealing with moisture release by dilution throughout residence may be very different than the capacity for dealing with pollutant releases through dilution. Home interiors have a large buffering capacity for moisture. This can be helpful but also can lead to problems with mold. Increasing general whole-home ventilation may work better for moisture than for pollutants, but increasing whole-home ventilation also may be more problematic when humidity of outdoor air is greater than indoor air. Research is needed to understand all of this.

This research question is especially important for cooking moisture management in extremely airtight (i.e., Passive House-style), high performance homes.

R8. Evaluate Potential for Effective Unvented Range Hood

There is significant demand for unvented, recirculating range hoods. There are some in the high performance home industry who are looking to minimize envelope penetrations and the vent for the kitchen range hood is a common target for omission. In existing single and multifamily homes that currently have no exhaust hood, it may be cost prohibitive to install a vented range hood. In high performance homes, builders seek to reduce envelope penetrations in every way possible, and a vented range hood is perceived as a penetration that can be eliminated through the use of a recirculating range hood. In all these cases, there is a need for a recirculating range hood that is capable of effectively filtering particles and pollutants of concern (such as acrolein, nitrogen dioxide, carbon monoxide, formaldehyde and others). There are reasons to believe that currently available recirculating range hoods are not effective at filtering cooking pollutants, though this has yet to be verified through careful experimental study. Higher quality recirculating range hoods with more robust filtration systems are now beginning to come to market, but their efficacy has not yet been determined.

There is a research need to evaluate the efficacy of these more robust recirculating devices, which may constitute a new category of kitchen ventilation/filtration product. To ensure cooking moisture removal, these hoods may need to be combined with increased whole-home ventilation.

Technology and Product Development Roadmap (TPD)

T1. Develop Ashrae 62.2-Compliant Microwave Range Hoods

Over the range (OTR) microwave and range hood combination appliances are popular for their cost- and space-savings. Data reported by Klug et al. (Klug et al. 2011a) indicate that a large percentage of all homes and an even larger percentage of newer homes in California have OTR microwaves. Unpublished data offered by a meeting participant indicates a similar trend nationally. Currently there are few if any microwave range hoods that have airflows and sound ratings certified by HVI and no products that meet either ASHRAE or ENERGY STAR requirements for airflow and sound. This product gap represents a core challenge for high performance new home programs – including ENERGY STAR for Homes, since many homebuyers desire the OTR microwave. This is by far the most urgent product development need. ENERGY STAR's current workaround solution is to allow range hoods that do not have an HVI certified flow provided that a 6-inch smooth duct is installed to ensure low pressure drop, which bolsters effectiveness when the devices are used, and to minimize noise, which helps reduce one barrier to the devices being used.

T2. Expanded Selection of Quiet, Effective, Affordable and Energy-Efficient Range Hoods

The assessment of this gap is based on an internet-based review of available products and pricing and estimation of effectiveness based on form (dimension, shape) in relation to LBNL published test results for capture efficiency. At present, ASHRAE 62.2 compliance and ENERGY STAR certification can be regarded as an indication of quietness and energy efficiency, if not reliable effectiveness.

T3. Establish Automatic Range Hood Shut-off to Reduce Hazard Associated with Natural Draft Combustion Appliance Backdrafting

In many cases, range hoods are contributing to the failures of worst-case combustion safety tests. The development of an automatic shut off feature for hoods operating at airflows that could cause failures would have the potential to ameliorate this issue. Problems associated with worst-case depressurization testing occur as total available airflows increase and air tightness decreases. The current standard "worst-case" depressurization test involves measuring depressurization with all exhaust fans operating on their highest settings.

A timer on the range hood – which is often the exhaust fan with largest capacity -- would eliminate the possibility of sustained spillage caused by depressurization. The automatic duration should be roughly 60 minutes or less. This would also have a small energy benefit to avoid hoods being left on for hours.

T4. Integration of Kitchen Exhaust into Smart Ventilation Systems for Residences

Increasingly, building codes are requiring whole-home ventilation per standards such as ASHRAE 62.2. Integrated point-source and whole-home ventilation systems can account for ventilation provided by kitchen range hoods and bathroom exhausts and only provide whole-home ventilation for the remainder, maintaining indoor environmental quality while reducing the ventilation-related energy penalty. This integrated ventilation technology has been demonstrated in prototype form as the Residential Integrated VEntilation Controller (RIVEC)[34] and a similar integrated local and whole-home ventilation system is now commercially available[35].

T5. Ensure Performance with on-Board Diagnostics

Commissioning range hoods is notoriously difficult. The addition of accurate, reliable on-board diagnostic sensors for range hoods would facilitate both commissioning at the time of installation and ongoing monitoring and insurance of performance. Relevant performance metrics that the sensors could monitor include airflow, static pressure, and power consumption. Analog pressure taps could be included in addition to digital sensors. These sensors could be used in lieu of commissioning or used by a technician to diagnose a performance problem. To prevent callbacks, the sensors and their status would likely be hidden from the end user. Automobile on-board diagnostics could be a model for this.

T6. Develop Effective Range Hoods with Automatic Operation

Range hoods that operate automatically as needed have the potential to significantly reduce health impacts caused by exposure to cooking pollutants. Automatic operation features must be certified to pass a yet-to-be-developed standard test. Potential automation mechanism options include heat or humidity sensor, particle counter or direct wired or wireless tie-in to range controls. An outstanding question related to these automation methods is whether or not every automated range hood must be compatible with every range, or if it is permissible for manufacturers to develop automated range hoods that operate exclusively with their own ranges. And in general, will automated range hoods be compatible with only certain types of ranges? For example, will heat-sensor-controlled range hoods operate properly with induction cooktops? These kinds of compatibility questions need to be considered when developing standard methods of test and guidance for automated range hoods.

T7. Establish Effective Low-Energy Kitchen Ventilation Options

As other energy loads are reduced through efficiency, the energy associated with ventilation for indoor air quality will be increasingly scrutinized. Operating a high airflow vented range hood during every cooking event will effectively remove cooking pollutants, but this approach incurs fan and space conditioning energy penalties. Currently there are few very-low-energy options for the effective removal of cooking pollutants. To keep energy consumption low, kitchen ventilation will need to be designed as one component of an integrated home ventilation system. In addition, that system will need to be tailored to the particular home type and climate. A robust filtration recirculating range hood coupled with an ERV for whole-home ventilation may be cost-effective in the hot, humid conditions of the southeastern U.S., and for that region, such a system could be the best option for a very high performance home. By contrast, an induction cooktop coupled with a well-designed low-flow vented range hood may be a better solution in Los Angeles, where the space-conditioning penalty is much lower.

Filtration systems require regular cleaning and maintenance to remain effective, and the track record for user-maintenance of HVAC systems is poor. As such, recirculating range hoods' potential for effectiveness over time is anticipated to be compromised by their need to be regularly serviced.

For common home types and regions, demonstrations and evaluations of advanced kitchen ventilation systems in homes need to be developed. These systems could include technologies such as balanced ventilation with heat capture, low flow/high capture efficiency designs, or recirculating hood with effective filtration of relevant pollutants.

References

Abdullahi, K. L., Delgado-Saborit, J. M., & Harrison, R. M. (2013). Emissions and indoor concentrations of particulate matter and its specific chemical components from cooking: A review. *Atmospheric Environment, 71*, 260-294. doi: Doi 10.1016/J.Atmosenv.2013.01.061

Afshari, A., Matson, U., & Ekberg, L. E. (2005). Characterization of indoor sources of fine and ultrafine particles: a study conducted in a full-scale chamber. *Indoor Air, 15*(2), 141-150.

ASHRAE. (2010). ASHRAE Standard 62.2-2010: Ventilation and Acceptable Indoor Air Quality in Low-Rise Residential Buildings.

Belanger, K., Gent, J. F., Triche, E. W., Bracken, M. B., & Leaderer, B. P. (2006). Association of indoor nitrogen dioxide exposure with respiratory symptoms in children with asthma. *American Journal of Respiratory and Critical Care Medicine, 173*(3), 297-303. doi: 10.1164/rccm.200408-1123OC

Booth, B., & Betts, K. (2004). Cooking spews out ultrafine particles. *Environmental Science & Technology, 38*(8), 141a-142a.

BPI. (2012). Building Performance Institute Technical Standards for the Building Analyst Professional. Malta, NY.

Buonanno, G., Morawska, L., & Stabile, L. (2009). Particle emission factors during cooking activities. *Atmospheric Environment, 43*(20), 3235-3242. doi: Doi 10.1016/J.Atmosenv.2009.03.044

Delp, W. W., & Singer, B. C. (2012). Performance Assessment of U.S. Residential Cooking Exhaust Hoods. *Environmental Science and Technology, 2012*(46), 7.

Dennekamp, M., Howarth, S., Dick, C. A. J., Cherrie, J. W., Donaldson, K., & Seaton, A. (2001). Ultrafine particles and nitrogen oxides generated by gas and electric cooking. *Occupational and Environmental Medicine, 58*(8), 511-516.

Energy Star. (2013a). ENERGY STAR Certified Homes, Version 3 - Local Mechanical Exhaust Requirements for Kitchens.

Energy Star. (2013b). ENERGY STAR Certified Homes, Version 3 (Rev. 07) Inspection Checklists for National Program Requirements.

EPA. (2013). Indoor airPLUS Construction Specifications - Version 1, Revision 1.

Fortmann, R., Kariher, P., & Clayton, R. (2001). Indoor air quality: residential cooking exposures. Sacramento, CA: Prepared for California Air Resources Board.

Fugler, D. W. (1989). Canadian Research into the Installed Performance of Kitchen Exhuast Fans. *ASHRAE Transactions, 95(1)*, 6.

Fullana, A., Carbonell-Barrachina, A. A., & Sidhu, S. (2004). Volatile aldehyde emissions from heated cooking oils. *Journal of the Science of Food and Agriculture, 84*(15), 2015-2021. doi: 10.1002/jsfa.1904

Garrett, M. H., Hooper, M. A., Hooper, B. M., & Abramson, M. J. (1998). Respiratory symptoms in children and indoor exposure to nitrogen dioxide and gas stoves. *American Journal of Respiratory and Critical Care Medicine, 158*(3), 891-895.

Hansel, N. N., Breysse, P. N., McCormack, M. C., Matsui, E. C., Curtin-Brosnan, J., Williams, D. L., . . . Diette, G. B. (2008). A longitudinal study of indoor nitrogen dioxide levels and respiratory symptoms in inner-city children with asthma. *Environmental Health Perspectives, 116*(10), 1428-1432. doi: 10.1289/ehp.11349

He, C. R., Morawska, L. D., Hitchins, J., & Gilbert, D. (2004). Contribution from indoor sources to particle number and mass concentrations in residential houses. *Atmospheric Environment, 38*(21), 3405- 3415.

HVI. (2013). Certified Home Ventilating Products Directory: Certified Ratings in Air Delivery, Sound and Energy for Accurate Specifications and Comparisons - October 2013. In H. V. Institute (Ed.), (Vol. HVI Publications 911).

Klug, V. L., Lobscheid, A. B., & Singer, B. C. (2011). Cooking Appliance Use in California Homes – Data Collected from a Web-Based Survey *LBNL-5028E*. Berkeley, CA: Lawrence Berkeley National Laboratory.

Klug, V. L., Singer, B. C., Bedrosian, T., & D'Cruz, C. (2011). Characteristics of Range Hoods in California Homes – Data Collected from a Real Estate Web Site *LBNL-5067E*. Berkeley, CA: Lawrence Berkeley National Laboratory.

Less, B. (2012). *Indoor Air Quality in 24 California Residences Designed as High Performance Green Homes.* (Master of Science), University of California, Berkeley, Berkeley, CA. Retrieved from http://escholarship.org/uc/item/25x5j8w6#page-1

Li, Y., Delsante, A., & Symons, J. (1997). Residential kitchen range hoods - Buoyancy-capture principle and capture efficiency revisited. *Indoor Air, 7*(3), 151-157.

Li, Y. G., & Delsante, A. (1996). Derivation of capture efficiency of kitchen range hoods in a confined space. *Building and Environment, 31*(5), 461-468. doi: Doi 10.1016/0360-1323(96)00006-6

Lim, K., & Lee, C. (2008). A numerical study on the characteristics of flow field, temperature and concentration distribution according to changing the shape of separation plate of kitchen hood system. *Energy and Buildings, 40*, 9.

Madsen, U., Breum, N. O., & Nielsen, P. V. (1994). LOCAL EXHAUST VENTILATION - A NUMERICAL AND EXPERIMENTAL-STUDY OF CAPTURE EFFICIENCY. *Building and Environment, 29*(3), 319-323.

Moschandreas, D., Relwani, S., Johnson, D., & Billick, I. (1986). EMISSION RATES FROM UNVENTED GAS APPLIANCES. *Environment International, 12*(1-4), 247-254.

Moschandreas, D. J., & Relwani, S. M. (1989). Field-Measurements of NO2 Gas Range-Top Burner Emission Rates. *Environment International, 15*(1-6), 489-492.

Mullen, N. A., Li, J., & Singer, B. C. (2013). Participant Assisted Data Collection Methods in the California Helath Homes Indoor Air Quality Study of 2011-13 (Vol. LBNL-6374E). Berkeley, CA: Lawrence Berkeley National Laboratory.

Nagda, N. L., Koontz, M. D., Fortmann, R. C., & Billick, I. H. (1989). Prevalence, use, and effectiveness of range exhaust fans. *Environment International, 15*, 6.

Offermann, F. J. (2009). Ventilation and Indoor Air Quality in New Homes. Sacramento, CA: California Energy Commission and California Air Resources Board.

Parrott, K., Emmel, J., & Beamish, J. (2003). Use of Kitchen Ventilation: Impact on Indoor Air Quality. *The Forum for Family and Consumer Issues, 8*(1).

Piazza, T., Lee, R. H., Sherman, M., & P., P. (2007). Study of Ventilation Practices and Household Characteristics in New California Homes. Sacramento, CA: California Energy Commission and California Air Resources Board.

Rapp, V. H., Singer, B. C., Stratton, J. C., & Wray, C. P. (2012). Assessment of Literature Related to Combustion Appliance Venting Systems. Berkeley, CA: Lawrence Berkeley National Laboratory.

Rim, D., Wallace, L., Nabinger, S., & Persily, A. (2012). Reduction of exposure to ultrafine particles by kitchen exhaust hoods: The effects of exhaust flow rates, particle size, and burner position. *Science of the Total Environment, 432*, 350-356. doi: Doi 10.1016/J.Scitotenv.2012.06.015

Seaman, V. Y., Bennett, D. H., & Cahill, T. M. (2009). Indoor acrolein emission and decay rates resulting from domestic cooking events. *Atmospheric Environment, 43*(39), 6199-6204. doi: 10.1016/j.atmosenv.2009.08.043

See, S. W., & Balasubramanian, R. (2008). Chemical characteristics of fine particles emitted from different gas cooking methods. *Atmospheric Environment, 42*(39), 8852-8862. doi: 10.1016/j.atmosenv.2008.09.011

Singer, B. C., Apte, M. G., Black, D. R., Hotchi, T., Lucas, D., Lunden, M. M., . . . Sullivan, D. P. (2010). Natural Gas Variability in California: Environmental Impacts and Device Performance: Experimental Evaluation of Pollutant Emissions from Residential Appliances. Sacramento CA: California Energy Commission.

Singer, B. C., Delp, W. W., Apte, M., & Price, P. N. (2011). Performance of Installed Cooking Exhaust Devices *LBNL-5265E*. Berkeley, CA: Lawrence Berkeley National Laboratory.

Singer, B. C., Delp, W. W., Price, P. N., & Apte, M. G. (2012). Performance of Installed Cooking Exhaust Devices. *Indoor Air: International Journal of Indoor Environment and Health, 22*(3), 11.

Stratton, J. C., Walker, I. S., & Wray, C. P. (2012). Measuring Residential Ventilation System Airflows: Part 2 - Field Evaluation of Airflow Meter

Devices and System Flow Verification *LBNL-5982E.* Berkeley, CA: Lawrence Berkeley National Laboratory.

Wallace, L. A., Emmerich, S. J., & Howard-Reed, C. (2004). Source strengths of ultrafine and fine particles due to cooking with a gas stove. *Environmental Science & Technology, 38*(8), 2304-2311.

Zhang, Q. F., Gangupomu, R. H., Ramirez, D., & Zhu, Y. F. (2010). Measurement of Ultrafine Particles and Other Air Pollutants Emitted by Cooking Activities. *International Journal of Environmental Research and Public Health, 7*(4), 1744-1759. doi: 10.3390/ijerph7041744

APPENDIX 1: RESOURCES FOR BUILDING PROFESSIONALS

Better Homes and Gardens Kitchen Vent Guide

Provides non-technical guidance on choosing a range hood or downdraft kitchen exhaust, including considerations of installation and hood geometry.

Site:
http://www.bhg.com/kitchen/appliances/kitchen-vent-buying-planning-guide/

Building America Solution Center

Extensive guidance that incorporates Energy Star and ASHRAE 62.2

Site: http://basc.pnnl.gov/resource-guides/kitchen-exhaust

Home Ventilating Institute (HVI) Range Hood Brochure

Provides non-technical overview of the need for kitchen ventilation, what a range hood is, HVI's role in range hood quality assurance, and guidance on choosing a range hood, including HVI minimum and recommended airflow rates. Document: http://www.hvi.org/publications/pdfs/HVIRangeHood 4Feb08.pdf

Home Ventilating Institute (HVI) Tested/Certified Home Ventilating Fans Directory

September 2013. Site: http://www.hvi.org/proddirectory/

Document: http://www.hvi.org/proddirectory/sept2013/Sec1.pdf

LBNL RESAVE Ventilation Guide

Provides plain English guide to key elements of ASHRAE 62.2 and California Title 24, Part 6 with additional guidance and links.

Site: http://resaveguide.lbl.gov

APPENDIX 2: OBJECTIVES MATRIX

		Objectives											
		Expand awareness of the importance of kitchen ventilation.	**Expand use** of kitchen ventilation.	Publicize existing **guidance for building professionals** on the basics of kitchen ventilation and develop guidance on best practice designs and equipment specifications.	Provide better performance information and **guidance to purchasers** and users of kitchen ventilation products.	Improve and expand accessibility of **guidance for installers** and inspectors.	Raise awareness of kitchen ventilation **requirements of ASHRAE Standard 62.2.**	Establish **capture efficiency** as a key performance measure for range hoods.	Promote development of **automatic operation** as a future standard requirement.	Develop tools and guidance to **expand commissioning** of kitchen ventilation systems.	Reduce the number of new homes built without kitchen exhaust ventilation.	Incorporate kitchen ventilation as a standard element of energy **retrofits.**	Develop and demonstrate kitchen ventilation systems for low- and zero-netenergy, **high performance homes.**
Actions	**Make kitchen ventilation required**in the International Residential Code (IRC), International Mechanical Code (IMC), International Energy Conservation Code (IECC), International Green Construction Code (IgCC), Uniform Mechanical Code (UMC), and mechanical codes at state and local levels.	X	X			X				X	X		X

Appendix 2. (Continued)

		Objectives											
		Expand awareness of the importance of kitchen ventilation.	**Expand use** of kitchen ventilation.	Publicize existing **guidance for building professionals** on the basics of kitchen ventilation and develop guidance on best practice designs and equipment specifications.	Provide better performance information and **guidance to purchasers** and users of kitchen ventilation products.	Improve and expand accessibility of **guidance for installers** and inspectors.	Raise awareness of kitchen ventilation **requirements of ASHRAE Standard 62.2.**	Establish **capture efficiency** as a key performance measure for range hoods.	Promote development of **automatic operation** as a future standard requirement.	Develop tools and guidance to **expand commissioning** of kitchen ventilation systems.	Reduce the number of new homes built without kitchen exhaust ventilation.	Incorporate kitchen ventilation as a standard element of energy **retrofits.**	Develop and demonstrate kitchen ventilation systems for low- and zero-netenergy, **high performance homes.**
	Increase awareness of airflow rating certification; ensure that best practice guides and home performance standards **specify airflows certified at an appropriate static pressure**.	X		X	X	X	X						X

Objectives	Facilitate development and adoption of standard **method of test for capture efficiency** (pollutant removal effectiveness).	**Incorporate capture efficiency into standards** including ASHRAE 62.2, ENERGY STAR for range hood products, and high performance home standards.
Expand awareness of the importance of kitchen ventilation.		X
Expand use of kitchen ventilation.	X	
Publicize existing **guidance for building professionals** on the basics of kitchen ventilation and develop guidance on best practice designs and equipment specifications.		X
Provide better performance information and **guidance to purchasers** and users of kitchen ventilation products.	X	X
Improve and expand accessibility of **guidance for installers** and inspectors.		X
Raise awareness of kitchen ventilation **requirements of ASHRAE Standard 62.2.**		
Establish **capture efficiency** as a key performance measure for range hoods.	X	X
Promote development of **automatic operation** as a future standard requirement.		
Develop tools and guidance to **expand commissioning** of kitchen ventilation systems.		
Reduce the number of new homes built without kitchen exhaust ventilation.		
Incorporate kitchen ventilation as a standard element of energy **retrofits.**		
Develop and demonstrate kitchen ventilation systems for low- and zero-netenergy, **high performance homes.**	X	X

Appendix 2. (Continued)

		Objectives											
		Expand awareness of the importance of kitchen ventilation.	**Expand use** of kitchen ventilation.	Publicize existing **guidance for building professionals** on the basics of kitchen ventilation and develop guidance on best practice designs and equipment specifications.	Provide better performance information and **guidance to purchasers** and users of kitchen ventilation products.	Improve and expand accessibility of **guidance for installers** and inspectors.	Raise awareness of kitchen ventilation **requirements of ASHRAE Standard 62.2.**	Establish **capture efficiency** as a key performance measure for range hoods.	Promote development of **automatic operation** as a future standard requirement.	Develop tools and guidance to **expand commissioning** of kitchen ventilation systems.	Reduce the number of new homes built without kitchen exhaust ventilation.	Incorporate kitchen ventilation as a standard element of energy **retrofits.**	Develop and demonstrate kitchen ventilation systems for low- and zero-netenergy, **high performance homes.**
	Align Home Ventilating Institute (HVI) airflow test and sound procedures and ASHRAE standard 62.2 prescriptive compliance pathway to **achieve effective performance of installed range hoods**.			X	X	X	X						X

Research Topics	Objectives											
	Expand awareness of the importance of kitchen ventilation.	**Expand use** of kitchen ventilation.	Publicize existing **guidance for building professionals** on the basics of kitchen ventilation and develop guidance on best practice designs and equipment specifications.	Provide better performance information and **guidance to purchasers** and users of kitchen ventilation products.	Improve and expand accessibility of **guidance for installers** and inspectors.	Raise awareness of kitchen ventilation **requirements of ASHRAE Standard 62.2.**	Establish **capture efficiency** as a key performance measure for range hoods.	Promote development of **automatic operation** as a future standard requirement.	Develop tools and guidance to **expand commissioning** of kitchen ventilation systems.	Reduce the number of new homes built without kitchen exhaust ventilation.	Incorporate kitchen ventilation as a standard element of energy **retrofits.**	Develop and demonstrate kitchen ventilation systems for low- and zero-netenergy, **high performance homes.**
Incorporate automatic operation into best practice guides and high performance home standards as available products are determined to be reliable based on standard test method.		X		X				X				X
Incorporate into **combustion safety diagnostics** a consideration of whether the range hood has an automatic timer shut-off feature.	X	X	X						X		X	
Develop and demonstrate effective methods for commissioning range hoods.			X		X	X			X	X		

Appendix 2. (Continued)

		Objectives											
		Expand **awareness** of the importance of kitchen ventilation.	**Expand use** of kitchen ventilation.	Publicize existing **guidance for building professionals** on the basics of kitchen ventilation and develop guidance on best practice designs and equipment specifications.	Provide better performance information and **guidance to purchasers** and users of kitchen ventilation products.	Improve and expand accessibility of **guidance for installers** and inspectors.	Raise awareness of kitchen ventilation **requirements of ASHRAE Standard 62.2.**	Establish **capture efficiency** as a key performance measure for range hoods.	Promote development of **automatic operation** as a future standard requirement.	Develop tools and guidance to **expand commissioning** of kitchen ventilation systems.	Reduce the number of new homes built without kitchen exhaust ventilation.	Incorporate kitchen ventilation as a standard element of energy **retrofits.**	Develop and demonstrate kitchen ventilation systems for low- and zero-netenergy, **high performance homes.**
	Characterize in-field use and performance of range hoods.	X	X		X							X	
	Develop a comprehensive contaminant removal effectiveness metric.		X	X	X	X		X	X		X		X
	Improve or validate **ASHRAE 62.2 "deficit make-up" approach** for existing buildings.	X	X	X	X	X	X					X	

		Objectives											
		Expand awareness of the importance of kitchen ventilation.	**Expand use** of kitchen ventilation.	Publicize existing **guidance for building professionals** on the basics of kitchen ventilation and develop guidance on best practice designs and equipment specifications.	Provide better performance information and **guidance to purchasers** and users of kitchen ventilation products.	Improve and expand accessibility of **guidance for installers** and inspectors.	Raise awareness of kitchen ventilation **requirements of ASHRAE Standard 62.2.**	Establish **capture efficiency** as a key performance measure for range hoods.	Promote development of **automatic operation** as a future standard requirement.	Develop tools and guidance to **expand commissioning** of kitchen ventilation systems.	Reduce the number of new homes built without kitchen exhaust ventilation.	Incorporate kitchen ventilation as a standard element of energy **retrofits.**	Develop and demonstrate kitchen ventilation systems for low- and zero-netenergy, **high performance homes.**
	Evaluate reasonable combustion safety allowance for **timed auto-off kitchen ventilation.**	X	X	X	X	X						X	
	Develop standard test method and performance metrics for automatic operation.		X						X				X
	R7.Re-evaluate kitchen ventilation and moisture removal.	X		X	X	X							X
	R8. Evaluate potential for effective unvented range hood	X	X	X	X	X					X	X	X

Appendix 2. (Continued)

		Objectives											
		Expand awareness of the importance of kitchen ventilation.	**Expand use** of kitchen ventilation.	Publicize existing **guidance for building professionals** on the basics of kitchen ventilation and develop guidance on best practice designs and equipment specifications.	Provide better performance information and **guidance to purchasers** and users of kitchen ventilation products.	Improve and expand accessibility of **guidance for installers** and inspectors.	Raise awareness of kitchen ventilation **requirements of ASHRAE Standard 62.2.**	Establish **capture efficiency** as a key performance measure for range hoods.	Promote development of **automatic operation** as a future standard requirement.	Develop tools and guidance to **expand commissioning** of kitchen ventilation systems.	Reduce the number of new homes built without kitchen exhaust ventilation.	Incorporate kitchen ventilation as a standard element of energy **retrofits.**	Develop and demonstrate kitchen ventilation systems for low- and zero-netenergy, **high performance homes.**
Technologies	T1. Develop ASHRAE 62.2-compliant microwave range hoods	X	X	X	X	X					X		X
	T2.Expand the selection of quiet, effective, affordable and energy-efficient range hoods.		X		X						X	X	
	T3.Establish **automatic range hood shut-off** to reduce hazard associated with natural draft combustion appliance backdrafting.		X	X	X	X					X	X	

Objectives	Integration of kitchen exhaust into smart ventilation systems for residences.	Ensure performance through **on-board diagnostics**.	Develop effective range hoods with **automatic operation.**	Establish effective low-energy kitchen ventilation options.
Expand awareness of the importance of kitchen ventilation.		X		
Expand use of kitchen ventilation.			X	X
Publicize existing **guidance for building professionals** on the basics of kitchen ventilation and develop guidance on best practice designs and equipment specifications.	X	X	X	X
Provide better performance information and **guidance to purchasers** and users of kitchen ventilation products.	X		X	X
Improve and expand accessibility of **guidance for installers** and inspectors.	X	X	X	
Raise awareness of kitchen ventilation **requirements of ASHRAE Standard 62.2.**				
Establish **capture efficiency** as a key performance measure for range hoods.				X
Promote development of **automatic operation** as a future standard requirement.			X	X
Develop tools and guidance to **expand commissioning** of kitchen ventilation systems.		X		
Reduce the number of new homes built without kitchen exhaust ventilation.			X	X
Incorporate kitchen ventilation as a standard element of energy **retrofits.**				
Develop and demonstrate kitchen ventilation systems for low- and zero-netenergy, **high performance homes.**	X	X	X	X

End Notes

[1] California's 2008 energy code that first required kitchen ventilation did not come into effect until Jan 1, 2010.

[2] Thirty three percent had a shallow hood and 13% had a deep hood.

[3] Available at: (http://openpub.realread.com/rrserver/browser?title=/ASHRAE1/ashrae6222 0131024).

[4] To use the ASHRAE 62.2 prescriptive duct compliance pathway for new homes, the range hood fan flow must be rated by the Home Ventilating Institute procedures 915 and 916 at 62.5 Pascals (0.25 inches of water column) of static pressure.

[5] The fan airflow rate in cfm required to provide five kitchen ACH is equal to the kitchen volume in cubic feet divided by twelve.

[6] This is how ENERGY STAR for Homes defines kitchen area for the purposes of this calculation.

[7] Described in ASHRAE 62.2 (2010 and 2013 versions) Normative Appendix A – Existing Buildings

[8] http://www.energystar.gov/index.cfm?c=vent fans.pr crit vent fans

[9] http://www.hvi.org/ratings/Publication91609102013.pdf

[10] http://www.hvi.org/ratings/HVI915091013.pdf

[11] http://www.hvi.org/proddirectory/index.cfm

[12] http://www.hvi.org/publications/HowMuchVent.cfm

[13] http://publicecodes.cyberregs.com/icod/irc/2012/

[14] http://publicecodes.cyberregs.com/icod/iecc/2012/

[15] http://www1.eere.energy

[16] Description at footnote number 30 of the ENERGY STAR Certified Homes, Version 3 (Rev. 07) HVAC System Quality Installation Rater Checklist (http://www.energystar.gov /ia/partners/bldrslendersraters/downloads/InspectionChecklists.pdf?7e26-ed98)

[17] http://www.epa.gov/iaplus01/pdfs/constructionspecifications.pdf

[18] The most recent version of LEED for Homes was published in 2008, with updates in 2009 and 2010. http://www.usgbc.org/sites/default/files/LEED%20for%20Homes%20Rating%20 Systemupdated%20April%202013.pdf

[19] The kitchen ventilation requirements in ASHRAE 62.2-2013 are the same as those in ASHRAE 62.2-2007.

[20] http://www.usgbc.org/homes

[21] One simple modification that could improve the efficacy of this approach would be a control in the kitchen to boost ERV/HRV flows during cooking events.

[22] Capture efficiency expressed as an approximate percentage of the relevant pollutants captured by the range hood

[23] Total capacity of all burners and oven < 60 kBtu/hr (63.3 Mj/hr)

[24] The vast majority (92%) of the HVI range hood airflow ratings are at 25 Pa (0.10 IWC).

[25] Commonly referred to by the shorthand, "Title 24".

[26] Make-up air for >400 cfm range hood is required by IECC, IRC, IMC, and UMC.

[27] http://www.hvi.org/publications/pdfs/HVIRangeHood4Feb08.pdf

[28] The home ventilating fan directory is currently only available as a 282-page PDF file or a 7000+ row Microsoft Excel file

[29] http://www.hvi.org/proddirectory/sept2013/Sec1.pdf

[30] ASHRAE 62.2-2010

[31] For a more detailed discussion of the issues surrounding measurement of range hood airflows, see Stratton JC, Walker IS, Wray CP. 2012. Measuring Residential Ventilation System Airflows: Part 2 - Field Evaluation of Airflow Meter Devices and System Flow Verification. (LBNL-5982E). Berkeley, CA:Lawrence Berkeley National Laboratory..

[32] An accurate range hood airflow measurement device can be fashioned from an appropriately-sized cardboard box and a calibrated fan; however, this apparatus must be custom fabricated for each range hood and, despite its accuracy, has a makeshift appearance that may cause the homeowner to have reservations.

[33] This calculation procedure is described in more detail in Normative Appendix A – Existing Buildings of ASHRAE 62.2 – 2010. This calculation allows the use of exhaust fans airflow rated at 25 Pa (rather than 62.5 Pa), but the 25 Pa-rated airflows must be reduced by 25% for the purposes of the calculation.

[34] http://homes

[35] http://www.broan.com/media-

In: Kitchen Exhaust Ventilation
Editor: Mallory P. Michaels

ISBN: 978-1-63463-642-1

Chapter 2

PERFORMANCE OF INSTALLED COOKING EXHAUST DEVICES*

Brett C. Singer* [1], *William W. Delp, Michael G. Apte and Philip N. Price

ABSTRACT

The performance metrics of airflow, sound, and combustion product capture efficiency (CE) were measured for a convenience sample of fifteen cooking exhaust devices, as installed in residences. Results were analyzed to quantify the impact of various device- and installation-dependent parameters on CE. Measured maximum airflows were 70% or lower than values noted on product literature for 10 of the devices. Above-the-cooktop devices with flat bottom surfaces (no capture hood) – including exhaust fan/microwave combination appliances – were found to have much lower CE at similar flow rates, compared to devices with capture hoods. For almost all exhaust devices and especially for rear-mounted downdraft exhaust and microwaves, CE was substantially higher for back compared with front burner use. Flow rate, and the extent to which the exhaust device extends over the burners that are in use, also had a large effect on CE. A flow rate of 95 liters per second (200 cubic feet per minute) was necessary, but not sufficient, to attain capture efficiency in excess of 75% for the front burners. A-weighted sound

* This is an edited, reformatted and augmented version of a report, LBNL-5265E-r1(3), issued by the Lawrence Berkeley National Laboratory, November 2011.

levels in kitchens exceeded 56 dB when operating at the highest fan setting for all 14 devices evaluated for sound performance.

Keywords: Carbon monoxide; natural gas burners; nitrogen dioxide; range hood; task ventilation; unvented combustion

PRACTICAL IMPLICATIONS

Natural gas cooking burners and many cooking activities emit pollutants that can reach hazardous levels in homes. Venting range hoods and other cooking exhaust fans are thought to provide adequate protection when used. This study demonstrates that airflows of installed devices are often below advertised values and that less than half of the pollutants emitted by gas cooking burners are removed during many operational conditions. For many devices, achieving capture efficiencies that approach or exceed 75% requires operation at settings that produce prohibitive noise levels. While users can improve performance by preferentially using back burners, results suggest the need for improvements in hood designs to achieve high pollutant capture efficiencies at acceptable noise levels.

INTRODUCTION

Effective removal of air contaminants generated inside residences is needed to provide good indoor air quality and to protect the health and safety of occupants. Unvented gas appliances, including stovetops and ovens, can emit substantial quantities of carbon monoxide, nitrogen dioxide, formaldehyde and ultrafine particles (Dennekamp et al. 2001; Moschandreas and Relwani 1989; Singer et al. 2009; Traynor et al. 1996). Cooking activities (e.g. frying, grilling, and baking) contribute additional pollutants, odors and potentially undesirable levels of water vapor to indoor air (Fortmann et al. 2001). Cooking exhaust fans – including above the cooktop range hoods and downdraft systems – can remove burner exhaust and cooking-related contaminants before they mix throughout the home (Fugler 1989; Revzan 1986; Rim et al. 2011; Singer et al. 2011a; Singer et al. 2011b).

There are several relevant criteria for assessing cooking exhaust system performance. Primary among these is the efficiency at which contaminants generated on the cooktop and in the oven are exhausted before mixing into room air. Airflow, sound levels, and power consumption are also important performance parameters. While there is currently no rating system for capture efficiency of residential cooking exhaust devices, standard test methods and metrics exist for airflow, sound, and power consumption.

Airflow is expected from basic physical considerations to have a substantial impact on capture efficiency. The Home Ventilating Institute (HVI) certifies and publishes performance ratings for residential ventilation equipment, including cooking exhaust fans (HVI 2009a, 2010). HVI provides guidance on "minimum" (min.) and "recommended" (rec.) exhaust hood airflow rates in units of cubic feet per minute of airflow (cfm) per linear foot (lf) of cooking appliance width (HVI 2008). The guidance values are translated here to units of liters per second (lps) per linear meter (lm) using the conversion of 1.55 lps/lm per cfm/lf. The guidance values are 62 lps/lm min. and 155 lps/lm rec. for an appliance against a wall and 77 lps/lm min. and 232 lps/lm rec. for an island installation. For a U.S.-standard 30-inch (76.2 cm) wide range, these translate to 47 lps (100 cfm) min. and 118 lps (250 cfm) rec. for a wall installation and 59 lps (125 cfm) min., 177 lps (375 cfm) rec. for an island.

The level of sound produced during use is an important usability criterion. Though not definitive due to small sample sizes, restricted demographics, and self-reporting, the available data suggest that in the United States only about 10-25% of households always use their range hood while cooking, 40-60% use a hood sometimes, and 20-30% report never using their hood (Klug et al. 2011; Parrott et al. 2003; Price and Sherman 2006; Wilson et al. 1994). Excessive noise is one of the most cited reasons for avoiding routine use of cooking exhaust fans. HVI has a standard method for certifying sound levels (HVI 2009b).

In this work we investigate the extent to which cooking exhaust devices, if used, are effective at removing pollutants formed during combustion, and to what extent the effectiveness varies with airflow rate, device characteristics, and burner selection.

METHODS

On-site performance evaluations were conducted for an opportunity sample of 15 cooking exhaust systems installed in California residences. The sample included two downdraft exhaust units, two exhaust fan / microwave over-the-range (OTR) appliances, three installations of the same model of under-cabinet system with no substantial collection hood and grease screens covering the bottom inlet, and eight units with collection hoods. Comprising these eight were two island chimney units, one wall-mount chimney unit, and five under-cabinet units. The sample included three units installed at the time the kitchen was built, six units installed as part of a major kitchen remodel, two units installed by the current homeowners to replace a previously installed device, and four units installed by current homeowners into a kitchen that did not previously have an exhaust fan. Eight of the 13 above-the-cooktop devices were installed at heights consistent with product and/or industry recommendations, while others were slightly above or below the recommended height. The downdraft systems are incorporated into the cooking appliances. Details about experimental methods and photographs of the installed exhaust devices can be found in (Singer et al. 2011a).

Airflow

Airflow rates were measured using two approaches. Initially, airflow rates for F3, B2, and H2 were quantified by releasing sulfur hexafluoride tracer at a known rate into the hood intake and measuring concentrations downstream, and airflow rates for H2 and all other hoods were measured with a calibrated flow hood method described in detail by (Walker et al. 2001). With each method, the fan was operated through all settings, starting with the highest then stepping through to lower settings. The fan was not turned off between experiments in most cases. When laboratory experiments conducted subsequent to those described in this paper (Singer et al. 2011b) showed that microwave hoods may draw air from a second location above the microwave oven door, in addition to the primary pathway at the bottom of the microwave, we used the tracer method to (re-)measure total exhaust airflow through units F1 and F2.

Capture Efficiency

In the present paper, the primary metric for exhaust device effectiveness is "capture efficiency," used by (Li and Delsante 1996; Li et al. 1997) and others to quantify the fraction of generated pollutants that are pulled directly into the range hood before mixing into room air. Specifically, we quantify pollutant removal in terms of single-pass capture efficiency (SPCE), defined as the potentially time-varying fraction of pollutant mass emitted from the cooking appliance that is drawn directly into the exhaust system, i.e., before mixing into bulk room air.

The concentration of a pollutant P moving through the exhaust device, $C_{P,h(t)}$ (ppm or mL m^{-3}), can be measured directly and related to other parameters using the mass balance shown below:

$$Q_h\,[C_{P,h}(t) - C_{P,r}(t)] = \eta_D(t)\,E_p \qquad (1)$$

Here Q_h (m^3 min^{-1}) is the airflow rate through the exhaust system; $C_{P,r}(t)$ (mL m^{-3}, or ppm) is the concentration of P in the room air entering the diluted burner exhaust plume; $\eta_D(t)$ is the time-dependent capture efficiency; and E_P (mL min^{-1}) is the emission rate of P. The use of a time-varying room air concentration allows for an increase in this value resulting from emitted mass that is not captured on the first pass.

In the experiments conducted for this study, CO_2 was used as the marker for burner pollutants. The CO_2 mass emission rate was calculated from stoichiometry assuming complete combustion. The production rate of CO_2 (E_{CO2}) was calculated as follows:

$$E_{CO2} = Q_{fuel}\,N \qquad (2)$$

where E_{CO2} is the emission rate of CO_2 (mL h^{-1}), Q_{fuel} is fuel flow rate (mL h^{1}), and N is the molar fraction of carbon in the fuel (mol C per mol fuel), based on fuel composition. The fuel flow rate for each configuration of burners was checked using the home gas meter – taking care to subtract baseline fuel use for pilot lights, etc. – and when possible checked for consistency against the burner firing rates shown on the cooking appliance label. For all experiments, we used a value of N = 1.0246 for a representative Northern California fuel with 95% methane, 3% ethane, 0.2% propane, 0.9% carbon dioxide, 0.9% nitrogen (Singer et al., 2009). Based on our experience in conducting experiments with cooktops as described in Singer et al. (2009), we estimate that the incomplete combustion assumption and uncertainties in fuel flow measurement contribute <1% and up to about 5% respectively in the uncertainty of calculated capture efficiencies.

The concentration of CO_2 in the exhaust stream was measured with an EGM-4 infrared analyzer (PP Systems, ppsystems.com) that sampled either from the exhaust discharge or at a point as far downstream from the exhaust inlet as could be achieved. The analyzer has a rated accuracy of better than 1 percent of the span concentration over the calibrated range. The span calibration was checked at each experimental site with a verified standard mixture of 2,532 ppm CO_2. The concentration of CO_2 in room air was determined from measurements in the exhaust air before and just after each burner firing.

In all but one case, airflow measurements and capture efficiency experiments were conducted on the exhaust systems as found, i.e. without cleaning or replacement of grease screens. Three of the hoods had no inline grease screens by design, relying instead on impaction for grease collection. Among systems containing grease screens, most were lightly soiled with loading that did not appear to present a substantial incremental impairment to airflow. For exhaust system F2 (a microwave exhaust fan), the grease screen was found with a thick mat of accumulated grease and dust. The grease screen for this system was cleaned using dish soap and hot water and replaced in the hood before the experiment.

All cooktop experiments were conducted with covered 23-cm diameter, 5-L stainless steel pots filled with 3-4 L of room temperature water placed on each burner being used. The pots provided a physical barrier that somewhat cooled the plume of hot exhaust gases and forced the plume radially out from the center of each burner. Between experiments, pots were cooled to room temperature or slightly higher by exchanging water and through rotation. For oven experiments, the oven was set to 232 °C to ensure that the burner would operate for at least 5 min.

Capture efficiency experiments generally were conducted first at the highest fan setting, then at decreasing fan settings. The intent was to minimize buildup of CO_2 in room air resulting from incomplete capture. The exhaust fan generally remained on throughout each set of experiments.

Experiments were conducted with one of four burner configurations: oven only, single front cooktop burner, single rear cooktop burner, combination of one front and one rear cooktop burner. In almost all two-burner cases, the burners were diagonally opposed. In most homes, all four of these configurations were evaluated at both the lowest and highest fan settings. Some configurations were evaluated at intermediate settings.

In most experiments, we attempted to minimize external drivers of airflow in the vicinity of the cooking appliance. Adjacent windows were closed and the researcher avoided activity nearby to the cooktop. In a few experiments, the field researcher explored the effect of human activity by walking up to the cooktop once per minute, standing and stirring the pot of water for 30 seconds, then walking away. Results of these experiments are presented in Singer et al. (2011a).

Sound

The "sone" is a unit of sound intensity that is intended to relate to the perceived loudness of a sound. Since sones cannot be readily measured in field experiments, we quantified a related quantity: frequency-weighted sound levels, using the so-called "A-weighting" (dB-A). This is a standard approach for monitoring environmental and industrial noise containing multiple frequencies. We measured dB(A) with an Extech 407736 digital sound meter. Measurements were made at a standard position in front of the range/oven and at another location likely to be occupied in the kitchen (e.g., at a table or a food preparation area). The front of range position was 150 cm above the kitchen floor and 25 cm from the front of the range. Sound levels at both locations were measured under background conditions and for one or more fan settings. Background sound levels were measured with the exhaust fan off and without other experimental equipment operating. For some exhaust fans, the sound measurements were made subsequent to airflow and capture efficiency experiments.

Device and Installation Characteristics

The evaluated exhaust systems are divided into 4 basic design types. The most substantial distinction is between downdraft and above the range systems. Above the range devices are subdivided into units that feature a true capture hood or inverted "bowl" (identified as B1-B6); devices that provide minimal or no capture volume, having a "flat" profile (identified as F1-F5) and "hybrid" devices with a substantial capture volume above grease screens that cover the bottom inlet (H1-H2). The grease screens at the bottom of the hybrid units can impede vertical airflow even when clean and certainly when coated with grease and dust. The same impediment can occur on "bowl" units that have air inlets covered with grease screens.

For marketing purposes, cooking exhaust devices are organized into the following categories: under-cabinet (largest and broadest category); microwave-exhaust combination units (typically marketed as microwaves but including ventilation systems); chimney – wall; chimney – island; and downdraft. In our study, devices D1 and D2 are variations of the same basic design and nameplate of downdraft exhaust units; both draw air from behind the rear burners. Devices F1 and F2 are exhaust fans built into over-the-range (OTR) microwave ovens. Devices F3-F5 are essentially the same model of under-cabinet device installed in different configurations: F3 has no cabinets on either side of it, F4 has a cabinet on one side, and F5 has a cabinet on one side and a refrigerator on the other. Devices B3 and B6 are mounted above island cooktops. All of the other units are mounted between cabinets, although H1 has some space between the unit and the cabinets on either side. Units B3, B5 and B6 all contain the same fan unit.

Table 1 provides information about the devices evaluated. The devices spanned a wide range of prices and nominal airflows. Same or comparable new versions of the evaluated exhaust devices span a retail price range of under \$100 to roughly \$3000. While the total cost of the microwave units is listed as \$300, this cost includes a microwave oven; the exhaust fan component of this appliance typically is of lower cost and quality. Additional information about the devices is available in Singer et al. (2011a).

Recommended installation height above the cooking surface varied from 33–53 cm (13–21 in.) for B4 to 76–91 cm (30–36 in.) specified for H2. In general, the larger open hoods (B3, B5, B6) and "hybrid" units with large collection hoods above the bottom grease screens allow higher installations. A higher installation has less potential to interfere with cooking. Microwave units balance the objectives of mitigating obtrusiveness (e.g. through limited projection over front burners) and convenient access to the microwave, which requires that the appliance not be too high above the range. Most above the range systems in our study were installed within the recommended height range or just slightly above it. The largest deviation between recommended and actual was observed for F3, with an installed height of 99 cm (39 in.) and a recommended range of 41–58 cm (16-23 in.).

Table 1 additionally provides information about the fraction of the cooktop covered by the exhaust appliance. For each device, "cooktop coverage" was assessed as the fraction of burner area – defined by the grates over each burner – that was covered by the exhaust device. Cooktop coverage reported in Table 1 is for all four cooktop burners. Coverage of over the range exhaust systems exceeded 75% of cooktop burner area for eight of the units, was in the range of 50-75% for three units and was <50% for two installations. Interestingly, the sample includes examples of the same exhaust hood installed at similar heights but with very poor (F4) and very good (F5) cooktop coverage.

Table 1. Characteristics of cooking exhaust fans evaluated in this study

Device Characteristics (NA = not available; NR = not relevant)							Installation	
ID	Age (y)	Est. Cost[a]	Nominal airflow (lps)[b]	Number of fan settings	Sound (sones)[b]	Rec. height (cm)[b]	Installed Height (cm)	Cooktop Coverage[c]
				Downdraft				
D1	10	$1,350	283 [e]	4	NA[d]	NR [d]	NR [d]	NR [d]
D2	15+	$1,350	283 [e]	3	NA[d]	NR [d]	NR [d]	NR [d]
			Flat (No capture hood)					
F1	4	$300 [f]	142	5	NA[d]	>36	57	<50%
F2	4	$300 [f]	142	5	NA[d]	>36	48	50-75%
F3	9	$275	118	3	3.5	41–58	99	50-75%
F4	6	$275	118	3	3.5	41–58	64	<50%
F5	5	$275	118	3	3.5	41–58	69	>75%
		Hybrid (Capture volume above bottom grease screens)						
H1	3	$825	359 [e]	4	4.5	66–76	81	>75%
H2	5	$450	170	3	NA[d]	76–91	76	>75%
				Open (Bowl)				
B1	15+	$75	90	2	6	46–61	58	50-75%
B2	15	$950	283 [e]	Variable [g]	NA[d]	64	61	>75%
B3	6	$2,900	260 [e]	1	6	76	69	>75%
B4	8	$300	170	Variable [g]	5.5	33–53	69	>75%
B5	3	$1,250	260 [e]	2	6.5	69	69	>75%
B6	6	$2,900	260 [e]	1	6	76	65	>100%

[a] Estimated retail prices based on review of web-based retailers conducted in late 2009.

[b] Nominal (maximum) airflow provided in product literature as cubic feet per minute; sound and recommended installation height (inches) based on product literature.

[c] Fraction of cooktop covered by range hood.

[d] NA = not available; NR = not relevant to product design.

[e] Fan free air delivery.

[f] F1 and F2 are combined exhaust and microwave over the range (OTR) appliances. The quoted cost includes the cost of the microwave. The exhaust device adds less than $100, based on comparison to the cost of a microwave alone.

[g] Knob or control without any set levels.

Seven of the eight open and hybrid systems had cooktop coverage in excess of 75%. In all cases of incomplete cooktop coverage the front burners were not completely under the exhaust device. In two cases (devices F1 and F4) even the rear burners were not completely covered; these exhaust systems were set flush with the rear wall and the cooking range was set several inches out from the wall.

Later discussion of the effect of coverage on capture efficiency considers the extent to which the exhaust device sits directly over the burners that were actually in use for each experiment. We use the term "in-use burner coverage" to distinguish this from "cooktop coverage." Most devices had full coverage over the rear burners, and therefore had good in-use burner coverage for experiments that used rear burners only.

Multivariate Analysis

The influences of airflow rate and other parameters on capture efficiency were analyzed using multivariate regression. Indicator variables were created for each of the modeled device types (Bowl, Hybrid, Flat). Indicator variables were also created for each burner combination (Rear only, Front and Rear, or Front Only) and for the degree of in-use burner coverage (<50%, 50- 75%, >75%). For instance, in a test that used both a back burner and a front burner, coverage was "good" if the device extended over the rear burner and more than half of the front burner, "fair" if it extended over the back burner and less than half of the front burner, and "poor" if it covered only the rear burner (or less). Coverage was good for all rear burner experiments.

RESULTS AND DISCUSSION

Airflow

Measured airflows at various fan settings are presented in Table 2. (Table 2 also includes sound measurement results that are discussed later.) Measured airflows ranged from 35 to 180 lps (74 to 382 cfm) at the highest fan settings. The airflow rates as well as the ratio of airflows across settings varied substantially among the units evaluated. Higher-cost units tended to have higher airflows. However, the modestly priced F2-F5 and B4 ($300 and under) all had flows in the range of 105-120 lps (230-250 cfm) and H2 ($450) had among the highest airflows, moving about 70 lps (150 cfm) on low and about 170 lps (360 cfm) on high setting. For some devices the different settings covered only a narrow range of flow rates.

Table 2. Airflow and Sound Measurements

Hood		Airflow (lps)[a]			Measured Sound (dB(A))[b]			
ID	Low	Med	High	Nominal	Off	Low	Med	High
D1	101	110	114	283	45	64	66	67
D2	NM	120	136	283	36	NM	55	58
F1	23	44	67	142	43	51	56	61
F2	102	109	120	142	49	63	66	66
F3	24	74	108	118	36	40	47	57
F4	40	NM	116	118	NM	NM	NM	NM
F5	42	77	117	118	34	42	53	63
H1	85	94	106	359	42	62	64	67
H2	72	111	170	170	44	53	56	66
B1	21	NM	35	90	45	73	NA	65
B2	75	78	85	283	43	71	69	68
B3	NM	NM	105	260	42	NA	NA	67
B4	42	97	120	170	41	45	60	65
B5	120	NM	148	260	37	67	NA	71
B6	NM	NM	180	260	34	NA	NA	62

[a] Values shown are means of all measurements recorded. For devices with more than 3 settings, airflow and sound were measured at highest, lowest and one intermediate setting. For fans with fewer than 3 settings, some entries are not applicable (NA). Airflow not measured (NM) for some cases. Under laboratory conditions, the measurement methods are accurate to within 2-3%. For the conditions of this field study, we estimate an accuracy of 5% or better.

[b] Sound was measured at the front of the cooktop. The values shown represent an average over at least 30 seconds of measurement, with precision of 1 dB(A) or better.

For example, device F2 was tested with four fan settings, from “Low” to “Boost”, but the flow rates for these settings only varied from 102 to 120 lps (217 cfm to 255 cfm).

The ratio of measured to nominal (nameplate or manufacturer specified) airflow at the highest fan setting ranged from a high of 100% to a low of 28%. Four of the six HVI-rated units (representing two models, as F3-F5 are all the same model) had maximum airflows within 90% of the nominal values. A third model had maximum airflow of 71% of nominal and the fourth model was below 50%. The lowest performing of the HVI-rated systems was an older (>15 years) economy hood (B1). Of the nine units that were not HVI-rated, only three had maximum airflows that were greater than 50% of the nominal values and two of these were high-end products costing above $1000. The low-cost F1 unit had low airflow even on its high setting, but the similarly inexpensive F2 had much higher airflow. The substantially more expensive B2 and B3 did not achieve expected (nominal) airflow values. The nominal airflows noted for D1 and D2 were taken from the product literature for the fans installed in these units; this value appears to represent fan free-air delivery and does not even account for the pressure drop that results when the fan is installed into the downdraft appliance.

Capture Efficiency

Figure 1 shows capture efficiency for all experiments conducted with each exhaust system. It is apparent from Figure 1 that capture efficiency varies widely across the range of installed devices and conditions evaluated. For many devices there are very clear differences in performance related to burner selection and airflow. As the vast majority of experiments were conducted without interfering with airflow around the hood area, these results should be considered as best case conditions for these particular devices. Individual capture efficiency data points are estimated to be within 15-20% (on a relative basis) of the true, average value for the condition. This includes estimated uncertainties of up to 5% each for airflow, fuel flow, and incremental CO_2 measurements with additional but unevaluated uncertainty resulting from real trial-to-trial variability in exhaust hood airflow and capture efficiency performance.

Figure 2 presents the same data in a manner designed to display variations in capture efficiency by device type, airflow, burner selection, and the extent to which the device extends over the burners that are in use. Within each panel, capture efficiency is plotted against airflow. Panels show different combinations of burner configurations (columns) and device type (rows). Plotting symbols are the same as in Figure 1. Colors indicate different categories of in-use burner coverage. Dashed lines in the plots show predictions from a linear model that was fit to data from all of the devices except the downdraft types.

We used multivariate linear regression analysis to investigate the extent to which capture efficiency can be predicted from various combinations of explanatory variables. Predicting capture efficiency using linear regression is not correct on theoretical grounds. First, the efficiency can never be less than 0 or greater than 100%, but the prediction can.

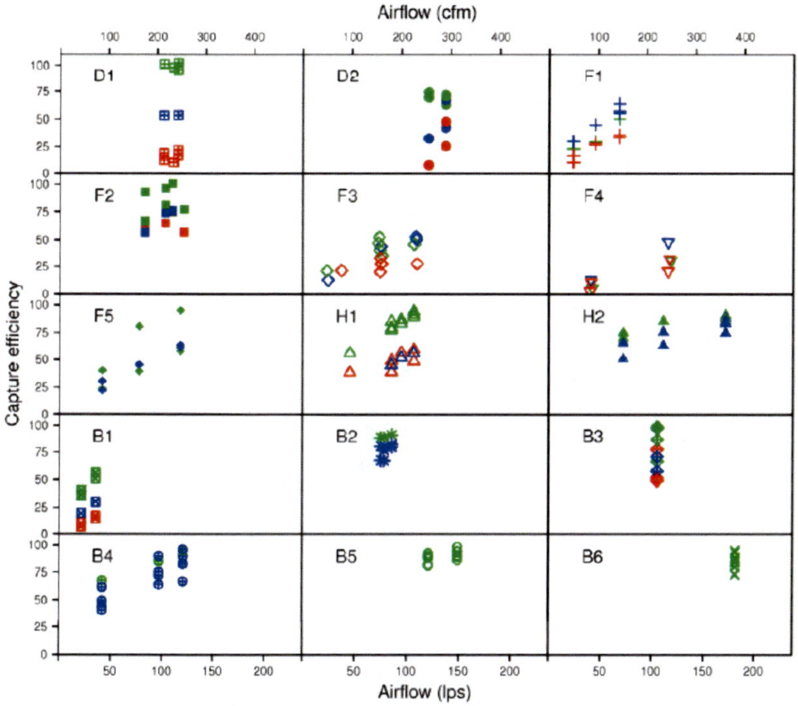

Figure 1. Single Pass Capture Efficiency for each device, for experiments using different fan settings and different combinations of burners. Colors indicate the degree to which the exhaust device extends over the burners that were used: green means >75%, blue means 50-75%, red means <50%. For Downdraft devices, green means a back burner was used, red means front, blue means both. Airflow is shown in liters per second on the lower axis and cubic feet per minute on the upper.

Second, the efficiency must go to zero as airflow goes to zero, but forcing this to be the case by using a no-intercept model generates poor fits. Alternatives to untransformed linear regression were considered and in some cases explored, but a nonlinear fit or a fit to transformed variables is much harder to interpret. Ordinary linear regression to predict capture efficiency yields easily interpretable results that fit well over the range of airflows observed in this study, but the results cannot be used to predict efficiency for very low or high flow rates. Also, the model results apply to the units as they were found, and cannot be used to predict how these units would behave if they were installed differently, such as with different configurations of adjacent cabinets and countertops.

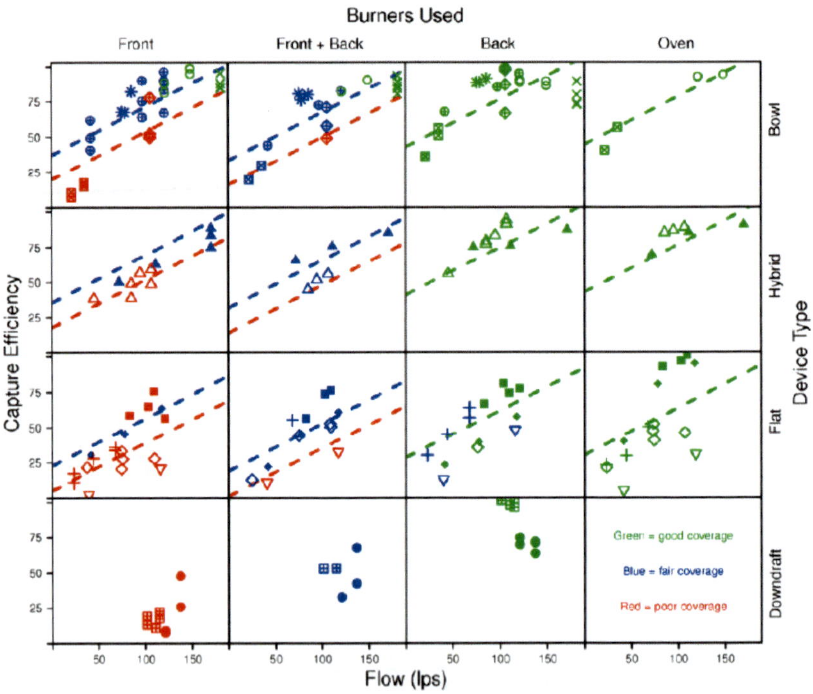

Figure 2. Capture Efficiency versus Flow (lps), for different device types (rows) and various combinations of burner or oven use (columns). The plotting symbols, a different symbol for each device, are the same as in Figure 1. Green symbols indicate high in-use burner coverage (>75% of the area of the burners in use is below the exhaust intake), blue indicates medium (50-75%), and red indicates low (<50%). For Downdraft devices, green means a back burner was used, red means front, blue means both. Dashed lines show predictions from model 7, for high, medium, and low in-use burner coverage. Predictions for high and medium coverage are almost identical, so those lines are nearly indistinguishable.

In some of the regressions, indicator variables were used for device type (bowl, hybrid, or flat) and/or for different combinations of burners in use. In each case, these were implemented as separate indicator variables for each type or combination. For instance, a "bowl" indicator variable took the value 0 for non-bowl devices and 1 for bowl-type devices.

Regression results are presented in Table 3. To illustrate the use of this table, consider using model 7 to predict the capture efficiency for a particular situation: a Flat device, operating at a flow rate of 106 lps (225 cfm), with one front burner in use, and poor coverage of the front burner. Referring to line 7 in Table 3, we calculate 28 + 33*1.06 -6-18=39, so model 7 predicts an efficiency of about 39% for this situation, with an uncertainty (one standard error) of 14 percentage points. In fact, two of the experiments met this general description: a test of device F3 at a flow of 109 lps (231 cfm) yielded a capture efficiency of 28%, and a test of device F4 at a flow of 115 lps (244 cfm) yielded a capture efficiency of 21%.

Table 3. Regression coefficients (standard errors), residual standard error (RSE), and R^2, for various models that predict capture efficiency (percentage points) based on the following variables: indicator for device type; flow in hundreds of lps; indicator for burner combinations (default is rear burner only); and indicator for extent of coverage over burners (default is >75%). Downdraft devices were not included

#	Inter-cept	Bowl	Hybrid	Flat	Flow (100s of lps)	Front and Rear	Front Only	Oven Only	In-use burner Coverage		RSE	R2
									Fair	Poor		
1		71 (2)	70 (4)	46 (3)							24	0.20
2	22 (3)				40 (2)						18	0.50
3	30 (4)				42 (2)	-12 (4)	-15 (3)	2 (4)			17	0.55
4	35 (4)				34 (2)				-7 (3)	-26 (3)	16	0.63
5		34	32	18 (3)	36 (3)						17	0.57
6		42 (3)	38 (4)	24 (3)	36 (3)	-11 (3)	-15 (3)	3 (3)			15	0.65
7		43	41	28 (4)	33 (3)	-9 (4)	-6 (5)	2 (4)	0 (4)	-18 (5)	14	0.69

Comparing models 6 and 7, we see that including indicator variables for the burners used, and including in-use burner coverage as well, yields only slightly better results (higher R^2, lower RSE) than including only the indicator variables for which burners were used, as long as device type is also included. However, the effect of in-use burner coverage is undeniably (and unsurprisingly) real, with poor coverage corresponding to an estimated 18-point drop in effectiveness, all other things being equal. To put that in perspective, an exhaust device with poor in-use burner coverage would have to provide an extra 52 lps (110 cfm) of airflow in order to be as effective as a similar type of device with fair or good coverage.

Line 1 of the table shows that Bowl and Hybrid devices performed about 25 percentage points better than Flat devices in removal efficiency; but, as Model 7 shows, the difference is reduced to about 15 percentage points when coverage and flow rate are taken into account. Flow rate is, as expected, very important, with each 50 lps (106 cfm) increase in flow corresponding to about a 15 percentage point increase in capture efficiency. All of the models summarized in Table 3 assume that the dependence of capture efficiency on flow rate is the same for all device types, as is the dependence on which burners were used, and on the coverage of the device.That is why there are only 9 coefficient estimates for Model 7, rather than the 56 that would be required to do a separate fit for each combination of device, airflow, and in-use burner coverage.

Different relationships surely hold for different device types, but with such small numbers of devices of each type it is not possible to get accurate parameter estimates by modeling each type separately. Model 7 predictions are displayed in Figure 2 for good coverage (green dashed line, which is uppermost), fair coverage (blue) and poor coverage (red, lower dashed line). Compare red lines to red symbols, blue to blue, and green to green. From the definition of model 7, all of the prediction lines have the same slope as a function of airflow, and all have the same downward offset associated with poor coverage; lines are shifted upwards or downwards according to the coefficient associated with the device (row) and the burners used (column).

Predictions for fair or poor coverage are not shown for back burners or oven, because all devices provide good coverage for these cases.

Model 7 explains about 70% of the variance in measured capture efficiencies. Sources of variation that are not included in the statistical models include: (1) device design features that are not captured simply by categorizing the devices into Bowl, Hybrid, or Flat; (2) installation features, such as whether adjacent appliances or walls help to guide air towards the fans; and (3) model misspecification, such as the fact that capture efficiency cannot in fact be linear in flow, as it is in the models.

Five of the thirteen non-downdraft devices behave substantially differently from the predictions from model 7: Devices B1 and B6 perform about 13 percentage points worse than predicted; F4 performs about 22 percentage points worse than predicted; B2 and F2 perform around 15 and 20 percentage points better than predicted. All of these differences are highly statistically significant; however, the prediction for B6 is for 100% efficiency so this device was guaranteed to underperform the prediction. For the other eight above the cooktop devices, the average capture efficiency is within 10 percentage points of the model predictions. The results for the anomalous devices other than B6 may be due to the situations in which they are installed. Photographs of the devices and their installation details can be found in Singer et al. (2011a). Device B1 and the stovetop under it are next to a doorway, not next to cabinets or at countertops as are most of the other devices; this may influence the airflow so as to decrease the amount of capture. In contrast, cabinets extend downwards towards the stovetop on both sides of device B2, which may help with capture. On the other hand, cabinets also extend downwards on both sides of H1, but that device performs only as well as the model predicts. Overall it appears that installation details, beyond simply the amount of coverage of the front burners, may promote or inhibit capture efficiency by as much as fifteen percentage points.

We also looked for systematic variation with the installation height – we tried using the height of the device over the cooking surface, and also height above the maximum recommended height as explanatory variables – but the resulting coefficient estimates were within one standard error of 0 and did not substantially improve the fits. The null finding for height is based on a limited sample of hoods and heights; height could be relevant for hoods or installation heights beyond those observed in this study. We also fit a variety of mixed-effects models (as described in (Gelman and Hill 2007)), with individual device coefficients as random effects, but the minor improvement in the predictions is not worth the added complexity of the models.

Performance degradation with equipment age could not be assessed with this small, disparate sample and likely varies considerably with device quality, cooking activity patterns and other factors. We note, however, that the 15 year-old B2 achieved capture efficiencies toward the top of the Bowl group despite having lower airflow than some other, newer devices in the group (Figures 1-2). This indicates that good performance is not necessarily compromised with age.

Sound

The primary goal of the present paper is to provide insight into the effectiveness with which kitchen exhaust devices remove pollutants generated during cooking, and the extent to which the effectiveness varies with features of the devices and the situations in which they are installed. However, any device is effective only when it is being used, and many people say noise is a deterrent to using their exhaust fan (Nagda et al. 1989; Parrott et al. 2003). We measured sound levels to document this important performance element.

Table 2 presents summary results for measured sound levels at the cooktop for 14 of the 15 installations. Sound levels elsewhere in the kitchen are presented in Singer et al. (2011a). Background sound levels varied from 34 to 49 dB(A) and increased to 57 to 71 dB(A) with exhaust fans operating on the highest settings. Even on medium or low fan settings many devices, coupled with the ambient noise already present, raised sound levels above 60 dB, a level that would strongly compete with normal conversation. Given these results it is not surprising that people are reluctant to use their exhaust devices.

The sound numbers in Table 2 show total measured sound levels, which includes both ambient noise and noise from the exhaust device. The sound level for the device alone can be determined by using the appropriate equation for subtracting sound levels (Davis and Davis 1997):

$$D = 10\log_{10}(10^{T/10} - 10^{A/10}) \tag{3}$$

where D is the noise level created by the device, A is the ambient noise level (i.e. the sound level measured with the the device turned off), and T is the total noise level when the device is turned on. Figure 3 shows device sound level (at the front of the cooktop) versus airflow for each measured airflow. Sound levels are extremely variable at low flow rates – some devices are very noisy even at quite low flow. Among the tested devices and fan settings that were effective at removing pollutants for all burner combinations, only H2 (filled triangles) on its medium fan setting was fairly quiet. Still, this device at least proves that a fairly quiet, effective device is possible. The extremely effective B6 device (X symbol) was fairly loud, but still quieter than many competing devices that moved much less air; potentially, a device like B6 operated at a lower airflow could be both quiet and at least moderately effective.

Product literature specified sound levels (in sones) for three fan settings of H1, for high and low fan settings of F3–F5 (variations of the same device), and for the highest fan settings of B1 and B3–B6. Among these devices and settings, the lowest specified sound levels were 0.9 sones for F3–F5 on low fan setting. This setting produced the lowest measured device-generated noise levels: 37 and 41 dB(A) for F3 and F5, respectively. The other devices and settings had sound specifications ranging from 2.0 to 6.0 sones and device-generated sound measurements of 57–71 dB(A) with an R^2=0.53.

It is important to note that the performance measurements reported in this study may reflect installation elements that were not assessed in this study. Improper device installation or ductwork can directly impact sound and airflow, and indirectly impact capture efficiency. The effect of such installation parameters is worthy of future study.

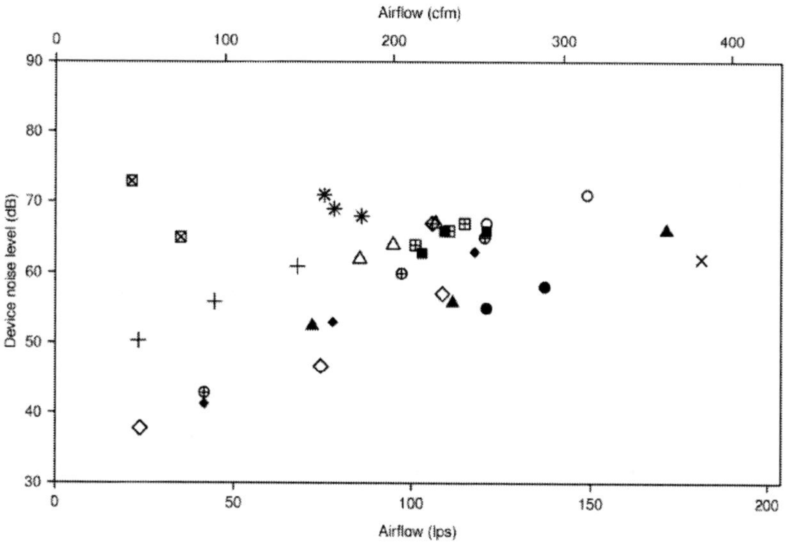

Figure 3. Sound increase over ambient levels versus airflow. A different symbol is used for each device, as indicated in Figure 1.

Conclusion

For any given device, higher airflow generally leads to higher capture efficiency, with each 100 lps increase in airflow increasing the capture efficiency by about 33 percentage points. However, at a given airflow there is a great deal of variation in capture efficiency among devices, and meeting industry standard guidance on minimum airflow requirements is not sufficient to ensure removal of most of the pollution generated by cooking.

The results of this limited study suggest that the most effective designs are those that include a capture volume and a robust fan. While the most effective devices (B5 and B6) are relatively expensive, the more moderately priced devices H2 and B4 also had high capture efficiencies under almost all conditions. B2 also performed well.

Whether the exhaust device extends over the burners being used also has a large influence: a device that does not cover the in-use burners suffers a very large penalty in capture efficiency, of 20-25 percentage points. As a consequence, for many of the under-cabinet and wall mount units (and for the back mounted downdraft systems), capture efficiencies were much higher on back as compared with front burners. In many cases, cooking on the rear burner with the fan on its low setting is as effective or more effective than cooking on the front burner with the fan on its medium setting.

The flat-bottom models in the study suffer from three problems: poor coverage of the front burners, low flow rates, and lack of physical features that would funnel pollutants towards the fan. The combined effect of these problems can be quite substantial: with the front burner in use, a flat device with poor coverage of the front burner and a weak fan can be expected to perform about 45 percentage points worse in removal efficiency than a hybrid or bowl device with good coverage and a fan that moves an extra 50 lps.

The devices with highest capture efficiency – F2, H2, B2, B4, B5, and B6 – remove more than 75% of cooking pollution when operated at medium fan setting or higher. To use these devices requires willingness to tolerate a moderately noisy kitchen, since all of these devices except H2 generate at least 60 dB of sound intensity at a medium fan setting or higher. Device H2 indicates that an effective, fairly quiet device is possible.

Acknowledgments

Funding was provided by the U.S. Department of Energy under Contract DE-AC02-05CH11231 and by the California Energy Commission through Contracts 500-05-026 and 500-08-061.

References

Davis, D. and Davis, C. (1997) Sound System Engineering, Second Edition. Massachusetts, Howard W Sams and Co., *Elsevier.*

Dennekamp, M., Howarth, S., Dick, C.A.J., Cherrie, J.W., Donaldson, K. and Seaton, A. (2001) Ultrafine particles and nitrogen oxides generated by gas and electric cooking, *Occupational and Environmental Medicine*, 58, 511-516.

Fortmann, R., Kariher, P. and Clayton, R. (2001) Indoor Air Quality: Residential Cooking Exposures, Final Report for California Air Resources Board Contract Number 97-330, Sacramento, CA.

Fugler, D.W. (1989) Canadian research into the installed performance of kitchen exhaust fans, *ASHRAE Transactions*, 95, 753-758.

Gelman, A. and Hill, J. (2007) Data Analysis using Regression and Multilevel/Hierarchical Models, New York, Cambridge University Press.

HVI. (2008) HVI Range Hood Brochure, Wauconda IL.

HVI. (2009a) HVI Airflow Test Procedure, Publication 916, 2009 Revised Edition, Wauconda IL, Home Ventilating Institute.

HVI. (2009b) HVI Loudness Testing and Rating Procedure, HVI Publication 915, Wauconda IL, Home Ventilating Institute.

HVI (2010) Certified Home Ventilating Products Directory, Wauconda IL, Home Ventilating Institute.

Klug, V.L, Lobscheid, A.B and Singer, B.C. (2011) Cooking Appliance Use in California Homes – Data Collected from a Web-Based Survey, LBNL-5028E, Berkeley CA, Lawrence Berkeley National Laboratory.

Li, Y. and Delsante, A. (1996) Derivation of capture efficiency of kitchen range hoods in a confined space, *Building and Environment*, 31, 461-468.

Li, Y., Delsante, A. and Symons, J. (1997) Residential kitchen range hoods - Buoyancy-capture principle and capture efficiency revisited, *Indoor Air*, 7, 151-157.

Moschandreas, D.J. and Relwani, S.M. (1989) Field-Measurements of NO2 Gas Range-Top Burner Emission Rates, *Environment International*, 15, 489-492.

Nagda, N.L., Koontz, M.D., Fortmann, R.C. and Billick, I.H. (1989) Prevalence, use, and effectiveness of range exhaust fans, *Environment International*, 15, 615-620.

Parrott, K., Emmel, J. and Beamish, J. (2003) Use of Kitchen Ventilation: Impact on Indoor Air Quality, The Forum for Family and Consumer Issues, 8, 1, Raleigh NC, North Carolina State University.

Price, P.N. and Sherman, M.H. (2006) Ventilation Behavior and Household Characteristics in New California Houses, LBNL-59620, Berkeley CA, Lawrence Berkeley National Laboratory.

Revzan, K.L. (1986) Effectiveness of local ventilation in removing simulated pollution from point sources, *Environment International*, 12, 449-459.

Rim, D., Wallace, L. and Persily, A. (2011) Reduction of exposure to ultrafine particles by kitchen exhaust fans of varying flow rates, In: Corsi, R.L. and Morrison, G.C., Eds., Proceedings of Indoor Air 2011, Austin TX, *International Society of Indoor Air Quality and Climate*, Paper 806.

Singer, B.C., Apte, M.G., Black, D.R., Hotchi, T., Lucas, D., Lunden, M.M., Mirer, A.M., Spears, M., Sullivan, D.P. (2009) Experimental Evaluation of Pollutant Emissions from Residential Appliances, CEC-500-2009-099, Sacramento CA, California Energy Commission.

Singer, B.C., Delp, W.W. and Apte, M.G. (2011a) Experimental Evaluation of Installed Cooking Exhaust Fan Performance, LBNL-4183E, Berkeley CA, Lawrence Berkeley National Laboratory.

Singer, B.C., Sherman, A.D., Hotchi, T. and Sullivan, D.P. (2011b) Pollutant removal efficiency of residential cooking exhaust hoods, In: Corsi, R.L. and Morrison, G.C., Eds., Proceedings of Indoor Air 2011, Austin TX, *International Society of Indoor Air Quality and Climate*, Paper 656.

Traynor, G.W., Apte, M.G. and Chang, G.M. (1996) Pollutant Emission Factors from Residential Natural Gas Appliances: A Literature Review, LBNL-38123, Berkeley CA, Lawrence Berkeley National Laboratory.

Walker, I.S., Wray, C.P., Dickerhoff, D.J. and Sherman, M.H. (2001) Evaluation of Flow Hood Measurements for Residential Register Flows, LBNL-47382, Berkeley CA, Lawrence Berkeley National Laboratory.

Wilson, A.L., Colome, S.D. and Tian, Y. (1994) California Indoor Air Quality Study, Irvine CA, *Integrated Environmental Services.*

End Note

[1] Corresponding Author: BCSinger@lbl.gov.

In: Kitchen Exhaust Ventilation
Editor: Mallory P. Michaels

ISBN: 978-1-63463-642-1

Chapter 3

POLLUTANT REMOVAL EFFICIENCY OF RESIDENTIAL COOKING EXHAUST HOODS*

Brett C. Singer, Alexander D. Sherman, Toshifumi Hotchi and Douglas P. Sullivan

SUMMARY

Capture efficiency (CE) of exhaust from a natural gas cooking range was quantified for three common designs of residential range hoods in laboratory experiments: (A) microwave exhaust combination; (B) short hood with grease-screen-covered air inlet at bottom; and (C) deep, open hood exhausting at top. Devices were evaluated at varying installation heights, at highest and lowest fan settings, and with the hood installed 15 cm away from back wall with intent to improve CE for front burners. Each configuration was evaluated for the oven and for three cooktop burner combinations (two back, two front, one front and one back). At highest fan settings and standard installation against the wall, Hoods A and C captured back cooktop burner exhaust at >90% and Hood B at >80%. In this configuration, CE for front burner exhaust was 73-78% for Hoods A and C but only 46-63% for Hood B. CEs followed similar patterns but were substantially lower on the lowest fan speed. Installing the hood away from the wall improved CE for oven and front burners on

---

* This is an edited, reformatted and augmented version of a report, LBNL-4902E, issued by the Lawrence Berkeley National Laboratory, June 2011.

Hood A at low speed, but substantially reduced CE for back burners for all hoods at low and high speed.

Keywords: Carbon monoxide, nitrogen dioxide, range hood, unvented combustion, ventilation

Implications

This study reinforces past findings that common range hood designs can effectively remove pollutants when operated on the highest available setting and cooking is preferentially done on back burners. Yet, the considerable noise generated at high fan speed operation impairs their utility. Under ideal lab conditions, lower and quieter fan settings still capture more than half of the burner exhaust when using back burners but less than 40% is captured from front burners. Capture of the cooking-related pollutants under real world conditions is likely less efficient. These results indicate opportunities to improve range hood designs.

Introduction

Indoor cooking activities on a stovetop or in an oven produce air pollutants that can reach high concentrations in homes, especially when using natural gas burners. Residential cooking exhaust fans can remove pollutants, odors, and water vapor before they mix into the air of the home, thus mitigating exposure and moisture condensation (Li et al., 1997). A recent study by Singer et al. (2010, 2011) measured in-situ performance characteristics – including airflow, sound and pollutant capture efficiency (CE) – of fifteen exhaust systems and analyzed the impact on CE of design, airflow, height, and positioning of the hood relative to burners. A key finding was that full coverage of burners substantially improved CE but that many installed hoods – including almost all under-cabinet models – did not cover front burners. This paper reports on laboratory experiments designed to study the effect of horizontal position (burner coverage), installation height, and airflow on CE for three common under-cabinet range hood designs. One hypothesis was that CEs of currently available hoods could be improved by offsetting installation from the back wall to cover more of the cooktop.

METHODS

Experiments were conducted in a simulated kitchen area within a larger laboratory. The area (shown in left panel of Figure 1) consisted of a back wall with simulated countertop and top cabinets installed on either side of an opening designed for a U.S. standard 30-inch (75 cm) standalone cooking range and under-cabinet range hood. Hoods were installed on steel strut that allowed variation in vertical and horizontal positioning. Exhaust from the hoods was vented upward with nominal 6 or 7 inch (15 or 18 cm) round ducting and connector suitable to each hood. A damper installed above the connector was set to provide 25 Pa static pressure at the highest airflow setting for each fan. Two standard residential gas ranges were used. The first – a range used for previous experiments – had two cooktop burners of 12.7 MJ/h and two of 9.7 MJ/h. The second range, purchased new when the oven of the first range stopped working, had one burner of 12.7 MJ/h and three of 10 MJ/h. To allow gas line connection, the range was 3.8 cm from the back wall.

Three exhaust hoods were selected to represent common designs, based on available models identified on manufacturer and retailer web sites, and an assessment of in-use prevalence in California residences through review of photographs of homes for sale listed on *Zillow.com.* The hoods are shown in Figure 1, and characteristics are provided in Table 1. The combined microwave exhaust appliance (A) appears to be the most common type in California homes built in the last 5 years. Device B is rated as compliant with sound requirements of the residential ventilation standard (62.2) of the American Society of Heating, Refrigerating and Air-conditioning Engineers. The shallow hood on this unit is similar to those featured on many lower cost (and generally noisier) hoods, though some have slanted inlets that may better utilize the capture volume. Hood C was selected for its large collection volume, fully open bottom inlet, and back-to-front depth – features expected to yield higher CE based on results from Singer et al. (2011). Hoods were purchased from online retailers and evaluated without any simulated aging or soiling of grease screens, i.e. under optimal conditions.

Inlet and outlet airflow for each hood was measured with a calibrated, pressure-controlled, variable-speed fan (Minneapolis Duct Blaster, www.energyconservatory.com) (Walker et al., 2001). For the inlet side, the fan was connected via a custom-fabricated transition and the fan speed was adjusted to achieve a neutral pressure. On the outlet side, the fan was attached using flexible aluminum ducting and operated to match the duct static pressure described above. In both cases, the fan speed was translated to air flow based on the flow calibration.

Figure 1. Exhaust hoods A, B, C shown from left to right; test area shown in left panel.

Hood performance was assessed using single-pass capture efficiency (CE), calculated as the ratio of incremental (above background) CO_2 mass flow through the hood divided by the CO_2 mass emission rate from the cooking burners (Eq. 1). Symbols are used as follows: Q_{exh} (m^3 min^{-1}) is the exhaust airflow; $CO_{2,exh}$ (mL m^{-3} or ppm) is the concentration of $CO_{2,exh}$ in hood exhaust, $CO_{2,bkg}$ is the level in room air, and E_{CO2} (mL min^{-1}) is the CO2 emission rate.

$$CE = Q_{exh} [CO_2,exh - CO_{2,bkg}] / E_{CO2} \qquad (1)$$

The CO2 emission rate was calculated assuming complete combustion of a typical Northern California fuel with 95% methane, 3% ethane, 0.2% propane, 0.9% CO2, 0.9% N2, and a heating value of 38.1 MJ m^{-3}. The CO2 emission rate is calculated with Equation 2; Q_{fuel} is the fuel volume flow rate (mL h^{-1}) and N is the molar fraction of carbon in the fuel (mol C per mol fuel), equal to 1.0246 for the fuel noted above (Singer et al., 2010).

$$E_{CO2} = Q_{fuel} N \qquad (2)$$

Carbon dioxide concentrations were measured with an EGM-4 analyzer (ppsystems.com) sampling from the center of the exhaust duct 68 or 73 cm above the hood with data logged at 5 sec intervals. Background CO2 was determined before and after burner operation. Exhaust CO2 was the steady concentration achieved during burner operation. The analyzer was calibrated at the start of each day at 2473, 1483 and 494 ppm.

Fuel flow rates were measured with an inline dry gas meter and a stopwatch. Fuel flow rates were measured prior to the first experiments with each range then infrequently, based on prior experience showing little variability for residential gas ranges. Measurement frequency increased after flow rates for the Kenmore range were observed to vary.

Table 1. Characteristics of hoods evaluated

ID	Make & model [a]	Cost [b]	Depth [c] [cm]	Fan levels	Nominal airflow [d] [cfm]	Nominal sound [e] [sone]	Recom. height [f] [cm]	Measured airflow [g] [cfm]
A	Panasonic NN-SD277	$310	38.1	4	420, 300, 160, 130	NL [h]	≥35	360, 275, 164, 140
B	Broan QT230	$109	44.5	Cont.	220 [i]	4.5 [j]	46-61	188, 87
C	Vent-a-hood PR9-130	$471	53.3	1	273 [i]	5.4	61-69	238

(a) Excludes color code. (b) Actual purchase price. (c) From back to front of appliance; inlets for hoods A-B extend only part of this distance. Hood C is open at bottom with grease collection via impaction at exhaust near top of hood. (d) Nominal airflow (cubic feet per min.) from product literature, based on vertical venting. (e) – (f) Nominal sound and recommended installation height from product specifications. (h) Not listed. (g) Exhaust airflow with damper set to 25 Pa static pressure at highest fan setting; Hood B at lowest and highest settings (i) Rating at 25 Pa static pressure. (j) Compliant with ASHRAE 62.2, sound level must be <3 sone at ≥100 cfm.

Each experiment started with the hood in operation at the specified fan setting. A covered 5-L stainless steel pot filled with 3-4.5 L of room temperature water was placed on the cooktop burners being used. Cooktop burners were set to maximum; the oven was set to 450 F. After starting burner(s), the operator remained moved away from the range. At 5 min, the operator approached and stirred the water in each pot for about 30 s total or opened the oven for 10 s, then walked away. This was done to explore the impact of cooking activity on capture efficiency. Burners were turned off at 7 min and data were collected for at least 1 more minute to establish the final background CO_2 level. For this paper, CE was calculated based on the steady mean concentration over the first 5 min.

Each exhaust system was evaluated at 3 different installed heights while flush against the wall and at 1-2 heights while extended 15 cm horizontally to improve coverage of front burners.

For the extended position, the gap behind the hood was sealed with a strip of wallboard and tape to avoid leakage of exhaust gases. Installation height was measured from the cooktop surface to the bottom edge of the exhaust appliance. The range of installed heights reflected manufacturer recommendations and judgment by our research team of likely installations, based on *Zillow.com* photos and a field study of installed exhaust fan performance (Singer et al., 2010). The microwave (A) was tested at heights of 35, 45 and 60 cm at the wall and 60 cm extended 15 cm from the wall. The other two hoods were tested at heights of 45, 60, and 75 cm at the wall and at 60 and 75 cm extended to cover burners. For each installation we evaluated a hood at the highest and lowest available fan flow settings for these burner combinations: two *back* burners, two *front* burners, highest firing rate *diagonal* combination of front and back burner, and *oven* (vents at back of cooktop).

Data quality was assessed through analysis of repeated conditions and by conducting several experiments with a custom hood designed to capture 100% of cooktop exhaust.

RESULTS

Measured exhaust airflows are presented in Table 1. Hood A exhaust airflows exceeded advertised values at the two lowest settings. Exhaust airflows were only 85-90% of advertised values for the two highest settings on Hood A, the high setting on Hood B and the single speed for Hood C. Flows measured at the inlet of Hood A were only about 67% of the flows measured in the exhaust duct for all fan settings. The extra airflow appears to be pulled from the microwave compartment. This result impacts the methods required to evaluate airflow and capture efficiency in residences where access to the exhaust side may be unavailable. We do not yet know if this design is common to microwave exhaust devices.

Calculated capture efficiencies (CE) are shown in Figure 2. Based on analysis of replicates (repeatability) and experiments conducted to verify overall method accuracy, any individual data point should be regarded as accurate to within ±10% of the value. Oven results are shown at right because the oven outlet directed exhausted from the back of the cooktop forward, with an effective release point assessed to be forward of the front burners.

Capture efficiency varied across the hoods, by airflow setting for each hood, by burners, and for some combinations of these factors, by installation location. For the two hoods with various fan settings, CE followed generally similar patterns across burner configurations and was in almost all cases substantially higher at the higher fan setting. Differences in CE between low and high fan settings were greatest for the front burner and oven for Hood A installed at the wall and for back burners for Hoods A and B installed away from the wall.

When installed flush to the back wall, all three hoods had substantially better CE for back burner versus front cooktop burner use. For one front and one back burner (diagonal), CE was generally between the values for two front or two back burners. Installed flush to the wall, CE for oven use was generally better than for front burners. At the highest fan settings and hoods flush to the wall, CE for back burners was at >80% for all three hoods, whereas CE for front burners varied from roughly 46-63% for Hood B to 73-78% for Hood A. At the lowest fan setting and hoods flush to the wall, CE for Hoods A and B was roughly 55-70% for back burners but only 20-40% for front cooktop and oven use. Still focusing on the panels at left of Fig. 2 (flush to the wall installations), the spread in CE seen for each hood, fan setting, and burner configuration is an indication of the impact of installation height. In almost all cases, install height caused less variability in CE than the other factors. The obvious exception is for oven use with Hood B on highest fan setting. For this operational condition the lowest installation (45 cm) achieved substantially better CE than the higher installation of 75 cm.

When installed 15 cm from the wall there was a small CE improvement for front burner with Hood C and substantial improvement for front and oven burners with Hood B on the lowest airflow setting. In contrast, this installation lowered CE for back burners on all evaluated airflow settings for all three hoods.

DISCUSSION

The results presented in this controlled laboratory study should be considered as best case. Airflows for installed devices may be lower due to poor duct design, blocked ducts and attic grills and other flow restrictions. Capture efficiency may be reduced when open windows and movements of the cook interfere with airflow patterns. The use of higher heat (e.g. "power") burners or more burners at one time is also expected to reduce capture efficiency.

Results presented in this paper are mostly consistent with results of recent performance measurements for a sample of 15 kitchen exhaust devices installed in residences (Singer et al., 2010, 2011).

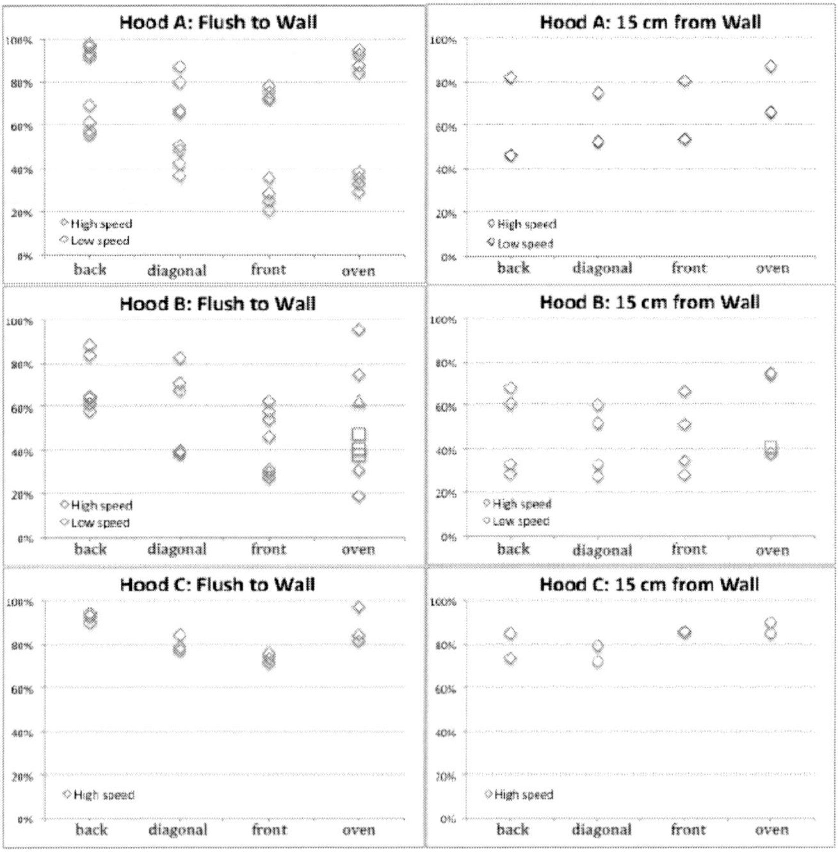

Figure 2. Capture efficiency by installation location and burners. Height distinguished only for Hood B oven experiments, with diamonds (45 cm), triangle (60 cm) and squares (75 cm).

A multivariate model developed from the data in that study suggested that design, airflow, burner choice and the extent to which the hood covered the burners being tested all had significant influences on CE. The finding in the laboratory experiments that the microwave hood had higher exhaust than inlet airflow raises questions about the calculated CEs for microwaves in the field study. Calculated CE is based on CO_2 mass flow in the exhaust, which is calculated as the product of exhaust CO_2 and airflow (Eq. 1). For hoods installed in homes, airflow was measured only on the inlet side. If exhaust airflow is substantially higher, than using the inlet airflow leads to an underestimate of CE. A tracer based approach – in which tracer gas is released at a known rate at a location that is fully captured by the hood, then measured in the exhaust - can be used to check the exhaust airflow from installed microwave hoods.

Preliminary results from screening photos of homes for sale on *Zillow.com* indicate that for new homes in California (built 2005-2010) the microwave over the range exhaust fans are the most common type. Results for Hood A suggest that these devices can be effective when cooking is preferentially done on back burners and the hood is operated at the highest airflow setting. The results of this study indicate that the lowest airflow settings – which are the most quiet and thus, assumed to be the most commonly used – fall far short of consistent, effective performance.

Conclusion

Exhaust pollutant capture efficiencies (CE) for common cooking exhaust devices vary substantially with airflow rate and device design, but not so much with installation height (range of 35 to 75 cm). In these and previous experiments, the best CE was achieved for cooking on back cooktop burners at airflow settings in excess of 200 cfm. Installing standard hoods farther from the wall does not consistently improve performance for front cooktop burners but does harm CE for back burner use. Information about hood performance would enable purchasers to select more effective models and might create market pressure to improve hood performance across the board.

Acknowledgment

Funding was provided by the U.S. Department of Energy under Contract No. DE-AC02- 05CH11231 and by the California Energy Commission through Contract 500-08-061.

References

Li Y, Delsante A, Symons J. 1997. Residential kitchen range hoods - Buoyancy-capture principle and capture efficiency revisited, *Indoor Air* 7(3), 151-157.

Singer BC, et al. 2009. Natural Gas Variability in California: Environmental Impacts and Device Performance: Experimental Evaluation of Pollutant Emissions from Residential Appliances. California Energy Commission Report CEC-500-2009-099. LBNL-2897E.

Singer BC, Delp WW, Apte MG. 2010. Experimental evaluation of installed cooking exhaust fan performance. LBNL-4183E. Lawrence Berkeley National Lab, Berkeley, CA.

Singer BC, Delp WW, Apte MG, Price PN. 2011. In-situ performance of residential cooking exhaust devices. *In preparation for submission to Indoor Air.*

Walker IS, Wray CP, Dickerhoff DJ, Sherman MH. 2001. Evaluation of flow hood measurements for residential register flows, LBNL-47382, Lawrence Berkeley National Lab, Berkeley CA.

INDEX

A

Above-the-cooktop devices, viii, 57
access, 22, 64, 84
accessibility, 4, 45, 46, 47, 48, 49, 50, 51, 52, 53
activated carbon, 21
age, 73
agencies, 20
air pollutants, vii, 1, 80
air quality, vii, 1, 2, 6, 7, 17, 18, 20, 21, 25, 27, 39, 41, 58
algorithm, 25
architects, 19
ASHRAE Standard 62.2, vii, 2, 4, 10, 23, 40, 45, 46, 47, 48, 49, 50, 51, 52, 53
assessment, 3, 38, 81
asthma, 7, 40, 41
asthma attacks, 7
authorities, 30
automation, 3, 25, 36, 39
Automobile, 39
awareness, vii, 1, 3, 4, 18, 19, 20, 23, 26, 27, 45, 46, 47, 48, 49, 50, 51, 52, 53

B

backlash, 25
barriers, vii, 1
base, 3, 19
breathing, 2
browser, 54
building code, vii, 1, 3, 9, 11, 17, 24, 26, 28, 34, 38

C

cabinets, 9, 11, 13, 64, 69, 72, 81
calibration, 62, 82
capture efficiency, vii, 4, 14, 15, 21, 22, 23, 24, 25, 27, 30, 36, 38, 40, 42, 45, 46, 47, 48, 49, 50, 51, 52, 53, 57, 59, 60, 61, 62, 63, 66, 68, 69, 70, 71, 72, 74, 75, 76, 77, 80, 82, 83, 84, 85, 87
capture hoods, viii, 57
carbon, 2, 7, 37, 58, 61, 82
carbon dioxide, 61
carbon monoxide, 2, 7, 37, 58
CEC, 77, 88
certification, 4, 14, 22, 38, 46
chemical, 40
children, 7, 40, 41
cleaning, 7, 17, 28, 34, 40, 62
climate(s), 36, 40
CO_2, 61, 62, 68, 82, 83, 86
code requirement, vii, 1, 9, 17
collaboration, 30
color, 83
combined effect, 76

combustion, vii, 2, 5, 27, 28, 32, 35, 38, 49, 51, 52, 57, 58, 59, 61, 80, 82
commercial, 19, 36
compatibility, 39
complexity, 73
compliance, 5, 10, 11, 13, 23, 24, 31, 33, 34, 38, 48, 54
composition, 61
condensation, 80
conditioning, 28, 35, 39, 81
configuration, viii, 61, 79, 85
consensus, 21, 30, 36
construction, 6
consumers, 17, 21, 29
consumption, 28, 39, 59
contaminant, 5, 25, 34, 50
conversations, 14, 23
cooking, vii, viii, 1, 2, 3, 6, 7, 10, 14, 15, 16, 17, 18, 19, 20, 21, 22, 24, 26, 28, 29, 30, 35, 36, 37, 39, 40, 41, 43, 44, 54, 57, 58, 59, 61, 62, 63, 64, 65, 73, 75, 76, 78, 79, 80, 81, 82, 83, 87, 88
cooking exhaust devices, vii, 57, 59, 63, 87, 88
cooking-related air pollutants, vii, 1
cooktop burner, viii, 2, 7, 19, 62, 64, 79, 81, 83, 85, 87
correlation, 15
cost, 10, 11, 27, 37, 40, 64, 65, 66, 68, 81
covering, 60

D

data collection, 10, 33
database, 29
decay, 43
deep, open hood exhausting at top, viii, 79
deficit, 5, 23, 35, 50
degradation, 73
demonstrations, 40
Department of Energy, 11, 76, 87
depth, 18, 21, 81
designers, 21
deviation, 64
diamonds, 86
distribution, 20, 42
draft, 5, 28, 52
drawing, 13

E

education, 19
emission, 36, 41, 43, 61, 82
energy, vii, 1, 4, 5, 6, 8, 11, 14, 25, 26, 27, 35, 36, 38, 39, 45, 46, 47, 48, 49, 50, 51, 52, 53, 54
energy conservation, 35
energy consumption, 8, 27, 36, 40
energy efficiency, 38
energy recovery, 14
enforcement, vii, 1, 3
environmental protection, 20
Environmental Protection Agency, 6, 13
environmental quality, 38
EPA, 13, 30, 31, 41
equipment, vii, 2, 3, 4, 6, 7, 8, 17, 20, 21, 22, 45, 46, 47, 48, 49, 50, 51, 52, 53, 59, 63, 73
evidence, 17, 18, 22
exchange rate, 23
exhaust devices, viii, 18, 57, 60, 64, 74, 84
exhaust fan/microwave combination appliances, viii, 57
exposure, vii, 1, 3, 15, 19, 21, 22, 23, 25, 26, 30, 35, 39, 40, 41, 43, 77, 80

F

fear, 32
filters, 18, 30
filtration, 7, 34, 37, 40
flex, 8
flow field, 42
flow rates, viii, 22, 24, 27, 43, 57, 66, 67, 69, 74, 76, 77, 82
food, 7, 19, 63
formaldehyde, 2, 7, 37, 58

G

geometry, 8, 15, 44
growth, 34
guidance, vii, 2, 3, 4, 11, 12, 13, 21, 22, 29, 30, 39, 44, 45, 46, 47, 48, 49, 50, 51, 52, 53, 59, 75

H

hazards, 20, 21
health, vii, 1, 2, 17, 20, 21, 26, 27, 32, 36, 39, 58
health problems, 27
height, 9, 13, 60, 64, 65, 73, 80, 83, 85, 87
homeowners, 18, 27, 60
homes, vii, 1, 2, 3, 4, 6, 7, 9, 10, 11, 13, 14, 15, 16, 17, 18, 19, 20, 21, 23, 24, 26, 27, 30, 32, 33, 34, 35, 36, 37, 40, 45, 46, 47, 48, 49, 50, 51, 52, 53, 54, 55, 58, 62, 80, 81, 86, 87
House, 14, 31, 34, 37
housing, 20, 34
Housing and Urban Development, 6
human, 62
human activity, 62
humidity, 16, 36, 39
hybrid, 31, 63, 64, 65, 70, 76
hypothesis, 18, 80

I

ID, 65, 67, 83
ideal, 21, 80
improvements, 58
incomplete combustion, 61
individuals, 20
induction, 28, 39, 40
industry, vii, 1, 2, 3, 6, 17, 19, 20, 22, 27, 30, 33, 37, 60, 75
initiation, 8
inspectors, 4, 22, 26, 33, 45, 46, 47, 48, 49, 50, 51, 52, 53
IRC, 4, 12, 13, 21, 28, 32, 45, 54
islands, 13
issues, 55

K

kitchen exhaust, vii, 1, 3, 4, 5, 6, 7, 9, 10, 11, 12, 13, 16, 17, 18, 21, 22, 23, 26, 27, 28, 29, 33, 35, 43, 44, 45, 46, 47, 48, 49, 50, 51, 52, 53, 73, 76, 77, 85
kitchen exhaust ventilation, vii, 1, 3, 6, 7, 9, 12, 16, 17, 18, 21, 22, 23, 26, 27, 29, 45, 46, 47, 48, 49, 50, 51, 52, 53
kitchen ventilation, vii, 1, 2, 3, 4, 5, 6, 7, 8, 9, 10, 11, 12, 13, 14, 16, 17, 18, 19, 20, 21, 23, 24, 25, 26, 27, 29, 30, 31, 32, 33, 34, 35, 36, 37, 40, 44, 45, 46, 47, 48, 49, 50, 51, 52, 53, 54

L

lead, 25, 36
leakage, 83
LEED, 14, 31, 54
linear model, 68
longitudinal study, 41

M

majority, 26, 28, 54, 68
management, 17, 37
marketing, 18, 63
mass, 41, 61, 82, 86
material surface, 34
materials, 18, 34
matter, 2, 40
measurement(s), 11, 13, 23, 25, 28, 33, 55, 61, 62, 63, 66, 67, 67, 68, 74, 88
mechanical ventilation, 12, 26
media, 55
meter, 59, 61, 63, 82
methodology, 30
Microsoft, 54
microwave exhaust combination, viii, 79
microwaves, viii, 3, 18, 37, 57, 63, 86

Minneapolis, 81
mixing, 24, 58, 60
models, 3, 15, 19, 24, 25, 30, 31, 36, 67, 71, 72, 73, 76, 80, 81, 87
modifications, 32
moisture, 2, 5, 6, 7, 8, 14, 15, 16, 17, 18, 21, 24, 28, 34, 36, 37, 51, 80
mold, 34, 36
morphology, 14, 15

N

natural gas, viii, 19, 28, 58, 79, 80
neutral, 81
nitrogen, 2, 7, 37, 40, 41, 58, 61, 76, 80
nitrogen dioxide, 2, 7, 37, 40, 41, 58, 80
null, 73

O

omission, 37
opportunities, 6, 27, 34, 80
over-the-range microwave, vii, 2

P

parameter estimates, 72
participants, 9, 16, 18, 25
pathways, 13, 34
penalties, 39
performance measurement, 74, 85
performance ratings, 30, 59
permit, 13
photographs, 60, 81
physical features, 76
policy, 6, 33
pollutants, vii, 1, 2, 3, 6, 7, 8, 14, 15, 17, 19, 21, 22, 25, 26, 27, 28, 30, 32, 34, 35, 36, 37, 39, 40, 54, 58, 59, 60, 61, 73, 74, 76, 80
pollution, 9, 75, 76, 77
polycyclic aromatic hydrocarbon, 2, 7
poor performance, 25
practice guides, 4, 5, 29, 31, 46, 49
preparation, 63, 88
primary function, 36
product design, 65
product performance, 20
professionals, 4, 18, 19, 21, 22, 26, 29, 45, 46, 47, 48, 49, 50, 51, 52, 53
propane, 61, 82
protection, 17, 18, 25, 28, 58
prototype, 39
public awareness, 6
public education, 19, 20
public health, 20

Q

quality assurance, 44

R

RDD, 32
real estate, 9
rear-mounted downdraft exhaust, viii, 57
recognition, 19
recommendations, vii, 1, 3, 19, 30, 31, 60, 84
recovery, 14
regression, 66, 68, 69
regression analysis, 68
relief, 27, 28
requirements, vii, 2, 3, 4, 9, 10, 12, 13, 14, 17, 18, 21, 23, 24, 26, 27, 30, 31, 33, 37, 45, 46, 47, 48, 49, 50, 51, 52, 53, 54, 75, 81
Residential, v, 4, 10, 12, 20, 26, 28, 30, 32, 34, 39, 40, 41, 42, 43, 45, 55, 76, 77, 78, 79, 80, 87, 88
resistance, 2, 7, 8, 9, 15, 28
resources, 21
restrictions, 16, 35, 85
retail, 64, 65
risk(s), vii, 1, 21, 32, 35
room temperature, 62, 83

S

safety, 2, 5, 27, 32, 33, 35, 38, 49, 51, 58
savings, 27, 37
sensors, 39
shape, 38, 42
short hood with grease-screen-covered air inlet, viii, 79
shortness of breath, 7
showing, 25, 83
simulation, 35
solution, 38, 40
specifications, 4, 12, 18, 45, 46, 47, 48, 49, 50, 51, 52, 53, 74, 83
stakeholders, vii, 1, 3, 6, 19, 20
standard error, 71, 73
state(s), 2, 4, 6, 9, 12, 17, 20, 24, 26, 28, 34, 45
statistics, 9
steel, 62, 81, 83
stoichiometry, 61
stoves, 41
structure, 20
style, 19, 37
sulfur, 60
symptoms, 7, 40, 41

T

target, 3, 37
technician, 25, 39
technology(s), vii, 1, 3, 4, 6, 17, 22, 25, 28, 39, 40
temperature, 42, 62
tension, 3
test procedure, 11, 15, 30
testing, 14, 15, 31, 38
time frame, 34
trade, 11
training, 6
transformation, 17
trial, 68

U

uniform, 26, 30, 33
United, 59
United States, 59
US locations, vii, 1

V

variables, 66, 68, 69, 70, 71, 73
variations, 64, 68, 74
ventilation equipment, vii, 2, 3, 8, 20, 59
volatile organic compounds, 2

W

walking, 62
Washington, 9
waste, 25
water, 7, 8, 11, 32, 34, 54, 58, 62, 80, 83
water heater, 32
water vapor, 7, 8, 34, 58, 80
web, vii, 1, 9, 18, 65, 81
web sites, 18, 81
wheeze, 7
windows, 62, 85

Y

yield, 81